Principles of Signals and Systems

Orhan Gazi

Principles of Signals and Systems

 Springer

Orhan Gazi
Electronic & Communication Engineering Department
Cankaya University
Ankara, Turkey

ISBN 978-3-031-17791-0   ISBN 978-3-031-17789-7   (eBook)
https://doi.org/10.1007/978-3-031-17789-7

© The Editor(s) (if applicable) and The Author(s), under exclusive license to Springer Nature Switzerland AG 2023
This work is subject to copyright. All rights are solely and exclusively licensed by the Publisher, whether the whole or part of the material is concerned, specifically the rights of translation, reprinting, reuse of illustrations, recitation, broadcasting, reproduction on microfilms or in any other physical way, and transmission or information storage and retrieval, electronic adaptation, computer software, or by similar or dissimilar methodology now known or hereafter developed.
The use of general descriptive names, registered names, trademarks, service marks, etc. in this publication does not imply, even in the absence of a specific statement, that such names are exempt from the relevant protective laws and regulations and therefore free for general use.
The publisher, the authors, and the editors are safe to assume that the advice and information in this book are believed to be true and accurate at the date of publication. Neither the publisher nor the authors or the editors give a warranty, expressed or implied, with respect to the material contained herein or for any errors or omissions that may have been made. The publisher remains neutral with regard to jurisdictional claims in published maps and institutional affiliations.

This Springer imprint is published by the registered company Springer Nature Switzerland AG
The registered company address is: Gewerbestrasse 11, 6330 Cham, Switzerland

Preface

The book contains the basic concepts of signals and systems and is written for undergraduate engineering students. The topics covered in this book can be taught in a one-semester course. The reader of this book should have a strong knowledge of calculus. Integration, differentiation, and summation operations taught in calculus must be very well known. During the writing of this book, we decided to rewrite some formulas instead of referring to the page where the formula has been defined. The main reason for this is that the reader could get distracted by turning away from the page they are on to read the referred formula. We tried to include as many solved examples as possible in each chapter of the book. The reader should study and understand these examples very well.

In Chap. 1, general definitions of signals and systems are provided, and we try to motivate the reader to the subject. In Chap. 2, definitions of the basic signal functions used in the literature are laid out. In Chap. 3, energy and power of the signals are explained. In Chap. 4, we discuss the Fourier analysis of continuous-time signals, one of the main topics of the book. The topics before Chap. 4 can be considered as preparatory subjects to the main topics. Fourier analyses of discrete-time signals, which are digital versions of continuous-time signals, are covered in Chap. 5. Laplace and Z-transforms, which can be considered as the generalized versions of Fourier analyses of continuous-time and discrete-time signals, are explained in Chaps. 6 and 7.

In Chap. 8, the practical applications illustrating the use of discrete Fourier transform are provided.

Balgat, Ankara, Turkey   Orhan Gazi

Contents

Abbreviations

CFTC	Complex Fourier Transform Coefficients
DTFSC	Discrete-Time Fourier Series Coefficients
FT	Fourier Transform
Im	Imaginary Part
LT	Laplace Transform
LTI	Linear and Time-Invariant
Re	Real Part
ROC	Region of Convergence
ZT	Z-Transform

Chapter 1
Introduction to Signals and Systems

A signal can be defined as a physical phenomenon that carries information. Examples of signals are human voice, electromagnetic waves, and sonar waves. Any quantity does not necessarily have to be in a waveform type to be considered as a signal. For example, Indians in ancient times used smoke to report any danger. The smoke used here is the signal carrier, and the shape of the smoke contains information. To give another example, let us say two persons are sitting in a room and one of these persons wants to describe an item to the other. For this, he only makes movements with his hands and tries to describe the item. We can accept the hands of the person as signal carriers and meaningful movements as signal. However, since movements containing meaning have to be with the carrier, in this case, we can accept the person's hands and the movements of the hands containing meaning as signal. Even a blink event and a facial expression can be given as an example of a signal. Signals change over time, and some are defined over a period of time. In order to show the changes of signals over time, it is necessary to describe them with some mathematical functions. The mathematical function used describes the changes in the signal over time and becomes a kind of synonym for the signal. For example, the mathematical expression

$$g(t) = \sin(2\pi ft) \quad 0 \le t < 1 \tag{1.1}$$

can be called both a sine function and a sine signal.

In this book, we will use the "signal" word and its corresponding mathematical function name interchangeably. Electromagnetic waves are one of the most common types of signals used today. Electromagnetic wave is used to transfer information from one place to another. The electromagnetic wave is the preferred type of signal carrier in the world of communication because it spreads with the speed of light and has the ability to pass through solid objects. Apart from this, laser or infrared signals are also used in short-distance communication systems.

© The Author(s), under exclusive license to Springer Nature Switzerland AG 2023
O. Gazi, *Principles of Signals and Systems*, https://doi.org/10.1007/978-3-031-17789-7_1

The electromagnetic wave has a sinusoidal graph of change. To send the information with this wave, we change the shape of the wave according to the information. For each different information, the wave takes a new shape. In this way, the information-loaded wave acquires the signal feature and is used for communication. Communication can be of two types which are analog communication and digital (digital) communication. The communication systems used in the past are based on analog communication. Today, digital communication systems are mostly preferred. In analog communication, the amplitude, frequency, and phase of the electromagnetic wave are changed according to the information wave, and the communication is performed. In digital communication, the continuous time data wave is first sampled. In other words, samples are taken from the data wave at certain time instances within a certain period of time. Each sample is expressed by a real number.

Real numbers are quantized to obtain integers, and their binary equivalents are calculated. For this purpose, a fixed number of bits can be used, for example, 8 bits. Later, groups of bits in the binary system are converted into data waves using square waves, analog modulation is performed, and the resulting wave is transmitted through an antenna. The steps of digital communication are shown in baseband in Fig. 1.1.

Human speech can also be given as an example of a signal. The voice of a person spreads in the atmosphere as sound waves and contains a certain state depending on the information we want to transmit. The form of the sound wave also changes as we move from situation to situation. For example, the sound wave produced by a singing person is different from the sound wave produced by a speaking person.

Another example of a signal is the current and potential difference values observed in circuits. By tracking voltage or current changes over a quantity, we gain information about that quantity. For example, the outputs of the devices connected to the heart to draw the beat graph of the human heart will generate voltage or current depending on the heart rhythm. We get information about the state of the heart from the graph of this voltage or current produced over time.

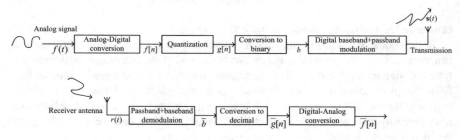

Fig. 1.1 Digital communication steps

1.1 System

As can be understood from the name of the system, a system refers to a kind of organization. This organization is created for a purpose. Systems receive data and produce outputs by processing the data they receive. Systems used in signal processing are generally electronic circuits. These electronic circuits receive signals as input data and generate signals as outputs. As an example, the electric circuit shown in Fig. 1.2 can be considered a system.

In Fig. 1.2, the input of the system is the voltage wave $v(t)$. The output of the system is the voltage $v_c(t)$ on the capacitor. Systems have certain features; we will examine these features in detail while explaining the system subject. In wireless communication, the signal is sent via an antenna and received via another antenna. The signal used in communication is electromagnetic wave. During the communication process, the data signal passes through many systems. For example, the power of the signal to be sent from the antenna is increased to higher energies by means of an amplifier.

An amplifier is nothing but an electronic system. Distortion will occur in the received signal at the receiving antenna. This distortion is caused by multipath communication and other factors in the environment, as well as thermal noise caused by the movement of electrons in the receiving antenna. To improve these distortions, the signal is passed through an equalizer. Equalizer is another electronic system. It receives a signal as input and generates another signal as output. All the sub-systems used in communication can be unified under the name "communication system."

Fig. 1.2 An electronic circuit can be considered as a system

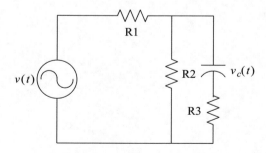

Chapter 2
Basic Signal Functions and Their Manipulations

Signals are physical quantities and they contain information. The changes of these physical quantities differ from each other. For example, the heart signals and the sound signals emitted from a musical instrument will differ from each other in some ways. Even so, many signals have certain common aspects when they are displayed graphically. For example, continuity in time can be given as an example of common aspects. We can generate complex signals through simple signals, just as we can construct a building using basic components. It is necessary to determine the main signals. Once we determine the main signals, we can generate more complex signals from these main signals. The first of the main signals is the sinusoidal signal. The sinusoidal signal can be obtained using a bipolar magnet and copper wire. The signal obtained in this way is an electrical signal. It is possible to obtain the DC voltage signal from the sine voltage signal by means of rectifiers. It is also possible to obtain square wave, impulse, and very narrow-spaced rectangular (impulse array) wave sequences by means of electrical circuits. An exponential voltage wave is observed during the discharge of the capacitor on the resistor. Once we have the fundamental waves, we can create other more complex waves. We can list the basic signals we use as follows: unit step signal, unit impulse signal, square wave signal, sinusoidal signal, exponential decaying signal, complex sine signal, etc.

There are some signals that do not have a smooth trend graph. The value of these signals at a certain moment of time cannot be calculated. Only a close guess can be made. These types of signals are called random signals. The amplitude of the signal received in multipath communication can be given as an example of random signals. In multipath communication, the amplitude of the received signal does not change properly depending on time. But the amplitude of the signal fits certain statistical distributions. This statistical distribution is expressed as a density function. The amplitude of the received signal in multipath communication is expressed by the Rayleigh density function if there is no direct line of sight between the transmitter and the receiver.

O. Gazi, *Principles of Signals and Systems*, https://doi.org/10.1007/978-3-031-17789-7_2

Signals are physical quantities that contain information. For literal expressions of these physical quantities, we express the signals with mathematical functions. We can divide the mathematical functions we use into two broad classes, mathematical functions with continuous variables and mathematical functions with discrete variables. Continuous variable signal naming can also be used instead of a continuous variable mathematical function. Similarly, instead of a mathematical function with a discrete variable, a signal with a discrete variable can be named. The variable parameter in signals is usually time or frequency.

Considering the time variable, the signals are grouped as either continuous-time signals or discrete-time signals. To inspect the signals in frequency domain, the Fourier transforms of the signals are evaluated. Fourier transform can also be called the spectral function of the signal. This function can be either continuous, discontinuous, or piecewise continuous on the frequency axis. Sometimes a single function is created that expresses the behavior of the signals in both time and frequency domains, and the change of the signal in these two planes can be observed with a single graph. Now, let us see the types of continuous and discontinuous signals. Then, let us look at how discrete-time signals are obtained by sampling from continuous-time signals.

2.1   Continuous-Time Signals

In continuous-time signals, the signal has a value for each moment of time, and the values the signal receives at the time t_i^- to t_i^+ are the same. The signal is continuous for a certain duration of time. The sine signal can be given as an example of continuous-time signals. Some signals may have discontinuities in certain places in their graphs. That is, the values of the signal at the time t_i^- to t_i^+ may be different. Such signals are called partial continuous-time signals. Square wave signal can be considered as a partial continuous signal. A signal can have discontinuity points in its graph, but this may not be sufficient for the signal to be called as discrete-time signal. Continuous-time and partial continuous-time signals are shown in Figs. 2.1 and 2.2.

Mathematical functions can be used for the representation of these signals, for example, the sine signal in Fig. 2.1 can be mathematically expressed as

$$f(t) = \sin(50\pi t)\ 0 \le t \le 0.1 \tag{2.1}$$

On the other hand, the pulse train signal shown in Fig. 2.2 must either be written in terms of known functions, or only one period of this periodic signal can be written mathematically. One period of wave in Fig. 2.2 can be written as

$$k(t) = \begin{cases} 1 & 0 \le t < 0.017 \\ -1 & 0.017 < t \le 0.034 \end{cases} \tag{2.2}$$

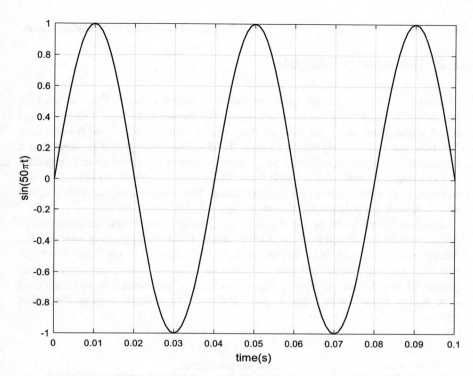

Fig. 2.1 Continuous-time sine function

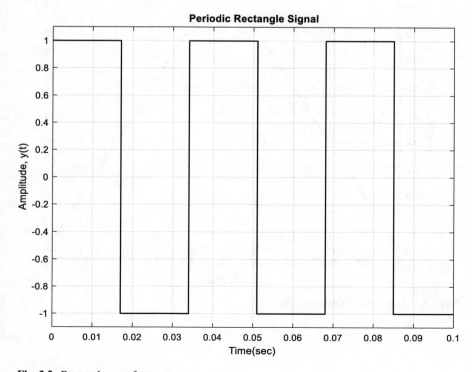

Fig. 2.2 Rectangle waveform

2.2   Discrete-Time Signals

Discrete-time signals acquire values at certain moments of time. They do not take any value in the interval between two successive moments of time. Other synonymous expressions used for discrete-time signals can be listed as discrete-time function, discrete-time signal function. If the discrete-time signal is expressed as a mathematical sequence, it should be specified which of the elements in this mathematical sequence indicates the start time. Discrete-time signals are obtained from continuous-time signals via sampling operation. The continuous-time triangle signal and its sampling operation are shown in Figs. 2.3 and 2.4.

Discrete-time signals can be displayed graphically or expressed mathematically. Digital signals are similar to discrete-time signals; however, in discrete-time signals, the time axis includes time information. Digital signals on the other hand are mathematical sequences, and time information is not explicitly available on the graphs of the signals. In digital signals, indices of the elements are considered time values. Some authors tend to use the discrete-time signal term as the digital signal.

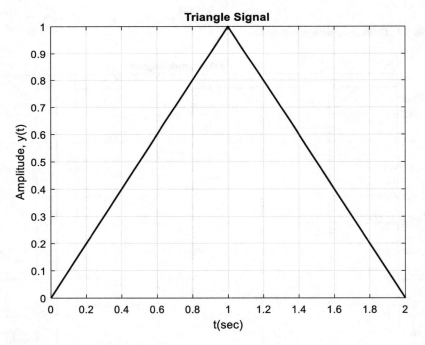

Fig. 2.3 Continuous-time triangle signal

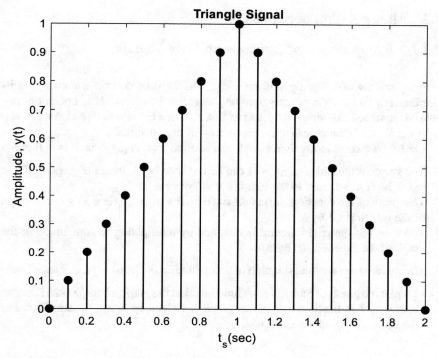

Fig. 2.4 Discrete-time triangle signal, $t_s \rightarrow$ Sampling time

Fig. 2.5 A digital signal

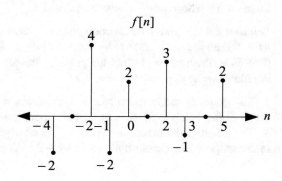

Example 2.1 Write the digital signal shown in Fig. 2.5 as a mathematical sequence.

Solution 2.1 $f[n] = \begin{bmatrix} -2 & 0 & 4 & -2 & \underbrace{2}_{n=0} & 0 & 3 & -1 & 0 & 2 \end{bmatrix}$

2.3   Manipulation of Signals

2.3.1   Manipulation of Continuous-Time Signals

New signals are obtained by shifting a signal in the time domain and changing its amplitude or taking its symmetry with respect to the time axis. These operations can also be explained mathematically. Let us first examine the processing of continuous-time signals and then consider the processing of digital signals.

Let $f(t)$ be a continuous-time signal, and assume that its graph is drawn, then:

1. The graph of the function $f(t - a)$ can be obtained by shifting the graph of $f(t)$ to the right, if $a > 0$, and to the left, if $a < 0$, by $|a|$ units.
2. The graph of the function $f(bt)$ is obtained by dividing the time axis of the graph of the function $f(t)$ by b.
3. The graph of the $cf(t)$ function is obtained by multiplying the amplitude of the graph of the function $f(t)$ by c.

There is also a combined situation of these three cases, let us explain it as follows:

4. To plot the graph of the $cf(bt - a)$ function, first the graph of the $f(t - a)$ function is drawn, then the time axis of this graph is divided by b, and the amplitudes of the obtained signal are multiplied by c.

Example 2.2 Plot the graph of the function $g(t) = -2f(2t + 3)$ obtained from the function $f(t)$ whose graph is given in Fig. 2.6.

Solution 2.2 To graph the function $g(t) = -2f(2t + 3)$ let us first graph the function $f(t + 3)$ then $f(2t + 3)$, and finally the function $-2f(2t + 3)$. The graph of the function $f(t + 3)$ is obtained by shifting the graph of the function $f(t)$ by 3 units to the left on the time axis as explained in Fig. 2.7.

The graph of the function $f(2t + 3)$ is obtained by dividing the time axis of the graph of the function $f(t + 3)$ by 2. This operation is explained in Fig. 2.8.

The graph of the function $-2f(2t + 3)$ is obtained by multiplying the amplitudes in the graph of the function $f(2t + 3)$ by -2 as illustrated in Fig. 2.9.

Fig. 2.6 A continuous-time signal

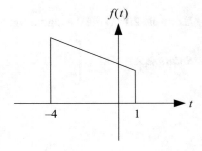

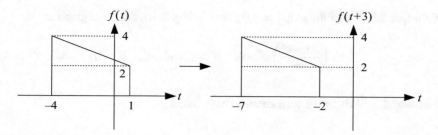

Fig. 2.7 Shifting a continuous-time signal

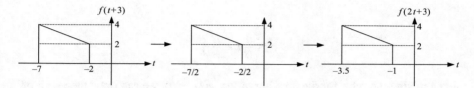

Fig. 2.8 Scaling of the shifted signal

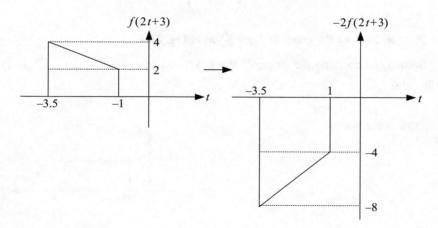

Fig. 2.9 Amplitude scaling on $f(2t + 3)$

Sometimes the mathematical expression of the processed signal is given as $c'f\left(\frac{t-a'}{b'}\right)$. To plot the signal $c'f\left(\frac{t-a'}{b'}\right)$ first the time axis of the graph of the $f(t)$ is multiplied by b' then the resulting graph is shifted to the right by a' (if $a' > 0$) or to the left by a' (if $a' < 0$), and finally the amplitude is multiplied by c'.

The more general form of $c'f\left(\frac{t-a'}{b'}\right)$ is written as $c'f\left(\frac{d't-a'}{b'}\right)$ whose graphing will be explained by an example.

Example 2.3 Using the graph of $f(t)$ given in Fig. 2.10, draw the graph of

$$c'f\left(\frac{d't-a'}{b'}\right) \quad c'>0, \quad b'>0, \quad a'>0, \quad d'>0 \tag{2.3}$$

Solution 2.3 Writing the mathematical expression

$$c'f\left(\frac{d't-a'}{b'}\right)$$

as

$$c'f\left(\frac{d'}{b'}t-\frac{a'}{b'}\right)$$

and applying the rules for the drawing of the $cf(bt-a)$, we can draw the graph of

$$c'f\left(\frac{d't-a'}{b'}\right)$$

First, let us draw the graph of $f\left(t-\frac{a'}{b'}\right)$ as in Fig. 2.11.

Second, let us draw the graph of the $f\left(\frac{d'}{b'}t-\frac{a'}{b'}\right)$ signal using $f\left(t-\frac{a'}{b'}\right)$ as in Fig. 2.12.

Fig. 2.10 A continuous-time signal

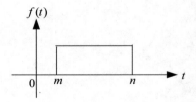

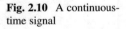

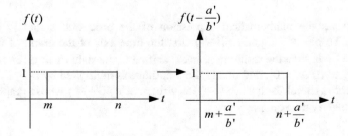

Fig. 2.11 Shifted signal

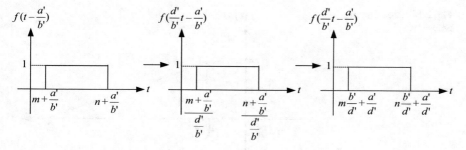

Fig. 2.12 Time scaling on the shifted signal

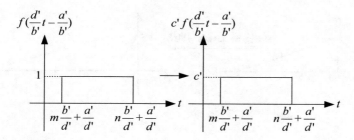

Fig. 2.13 Amplitude scaling on $f\left(\frac{d'}{b'}t - \frac{a'}{b'}\right)$

Lastly, we can draw the graph of $c'f\left(\frac{d'}{b'}t - \frac{a'}{b'}\right)$ is in Fig. 2.13.

Looking at the last graph, we see that to plot the graph $c'f\left(\frac{d't-a'}{b'}\right)$, the time axis of the graph of the $f(t)$ signal is multiplied by $\frac{b'}{d'}$ and shifted to the right by

$$\left|\frac{a'}{d'}\right|$$

if $\frac{a'}{d'} > 0$ and shifted to the left by

$$\left|\frac{a'}{d'}\right|$$

if $\frac{a'}{d'} < 0$.

Example 2.4 Using the graph of $f(t)$ given in Fig. 2.14, draw the graph of

$$f\left(\frac{2t-4}{5}\right) \tag{2.4}$$

Fig. 2.14 A continuous-time signal

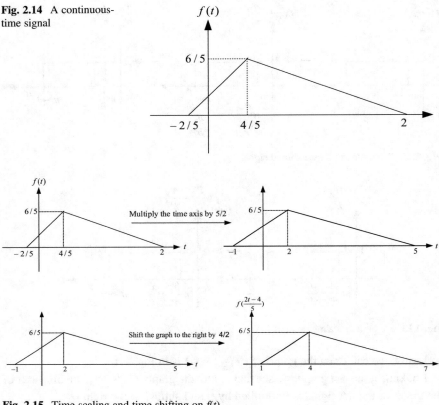

Fig. 2.15 Time scaling and time shifting on $f(t)$

Solution 2.4 To plot the $f\left(\frac{2t-4}{5}\right)$ signal, firstly multiply the horizontal axis of the graph of the $f(t)$ signal by $\frac{5}{2}$ and then shift the graph to the right by $\frac{4}{2}$ units as illustrated in Fig. 2.15.

Note In Solution 2.4, amplitude scaling was not needed because for this example $c' = 1$.

Example 2.5 The graph of the non-periodic continuous signal $X(w)$ with respect to the variable w is given in Fig. 2.16. Using Fig. 2.16, draw the graph of

$$Y(w) = \sum_{k} X\left(\frac{w - k2\pi}{T}\right) \tag{2.5}$$

Solution 2.5 Expanding the summation term in

$$Y(w) = \sum_{k} X\left(\frac{w - k2\pi}{T}\right)$$

we obtain

Fig. 2.16 Graph of the
continuous signal $X(w)$

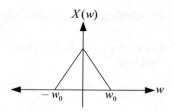

Fig. 2.17 Time scaling on
$X(w)$

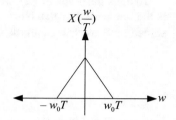

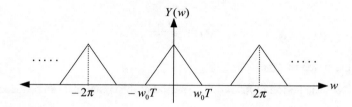

Fig. 2.18 Periodic signal $Y(w)$

$$Y(w) = \cdots X\left(\frac{w}{T} + \frac{4\pi}{T}\right) + X\left(\frac{w}{T} + \frac{2\pi}{T}\right) + X\left(\frac{w}{T}\right) + X\left(\frac{w}{T} - \frac{2\pi}{T}\right) + X\left(\frac{w}{T} - \frac{4\pi}{T}\right) + \cdots$$

First, let us draw the function $X\left(\frac{w}{T}\right)$ as in Fig. 2.17.

The graph of $X\left(\frac{w - k2\pi}{T}\right) = X\left(\frac{w}{T} - k\frac{2\pi}{T}\right), k \in Z$ is obtained by shifting the graph of $X\left(\frac{w}{T}\right)$ to the right or to the left by $|k2\pi|$ depending on the sign of k; for instance, for $k = 1$, we shift the graph to the right by 2π, and for $k = -1$, we shift the graph to the left by 2π.

Accordingly, the fundamental period of the $Y(w)$ function is 2π, and its graph can be drawn as in Fig. 2.18.

Manipulation of Periodic Signals (Functions)

Let the period of $f(t)$ be T, i.e. $f(t) = f(t+T)$ The period of the function $g(t) = cf(at+b)$ equals to

$$\frac{T}{a}.$$

To draw the graph of $g(t)$, we first draw the graph of $g_o(t) = cf_o(at+b)$ where $f_o(t)$ is the one period of $f(t)$. Next, shifting the graphs of $g_o(t)$ by $k\frac{T}{a}$ and summing the shifted replicas, we obtain the graph of $g(t)$ i.e., we obtain

$$g(t) = \sum_k g_o\left(t - k\frac{T}{a}\right)$$

Example 2.6 Using the periodic signal $f(t)$ depicted in Fig. 2.19, we define the function

$$g(t) = f\left(\frac{t-1}{2}\right) \tag{2.6}$$

Draw the graph of $g(t)$.

Solution 2.6 The period of $f(t)$ is 4. The period of the signal $g(t)$ can be calculated as

$$\frac{4}{1/2} \to 8$$

First, let us draw the signal

$$g_o(t) = f_o\left(\frac{t-1}{2}\right)$$

where $f_o(t)$ is the one period of $f(t)$, and then using $g_o(t)$, we obtain the periodic signal $g(t)$ as

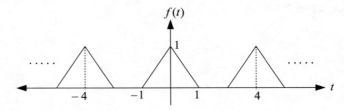

Fig. 2.19 Periodic signal $f(t)$.

Fig. 2.20 One period of $f(t)$

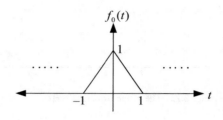

Fig. 2.21 The graph of $g_o(t)$

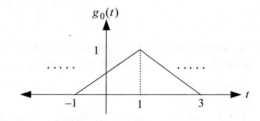

$$g(t)\sum_k g_o(t-k8)$$

where 8 is the period of $g(t)$. One period of $f(t)$, i.e., $f_o(t)$, can be chosen as in Fig. 2.20.

Shifting the graph of $f_o(t)$ to the right by $\frac{1}{2}$ and dividing the time axis of the shifted graph by $\frac{1}{2}$, we obtain the graph of

$$g_o(t)=f_o\left(\frac{t}{2}-\frac{1}{2}\right)$$

as depicted in Fig. 2.21. Or as an alternative approach, regarding the expression

$$g_o(t)=f_o\left(\frac{t-1}{2}\right)$$

we can first obtain the graph of

$$f_o\left(\frac{t}{2}\right)$$

by dividing the time axis of $f_o(t)$ by $\frac{1}{2}$ and then shifting the graph of $f_o\left(\frac{t}{2}\right)$ to the right by 1 unit, we can get the graph of

$$g_o(t)=f_o\left(\frac{t-1}{2}\right)$$

To get the graph of

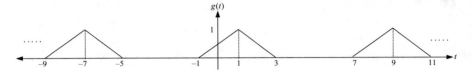

Fig. 2.22 The graph of periodic $g(t)$.

$$g(t) = \sum_k g_o(t - k8)$$

we shift the graph of $g_o(t)$ to the right and to the left by multiples of 8 and sum the shifted replicas as illustrated in Fig. 2.22.

2.3.2 Manipulation of Digital Signals

By shifting and scaling the digital signal $f[n]$, we can get $g[n] = af[bn - c]$. To get $af[bn - c]$, we perform the steps:

1. First, shifting $f[n]$ on the time axis by $|c|$, to the right or to the left (if $c > 0$, to the right, if $c < 0$, to the left), we get the graph of $f[n - c]$.
2. In the second step, the time axis of $f[n - c]$ is divided by b, and from those division results, only the integer division results are kept, the others are discarded, and in this way, we obtain the graph of $f[bn - c]$.
3. In the last step, the amplitudes of $f[bn - c]$, are multiplied by a, and the graph of $af[bn - c]$ is obtained.

Example 2.7 The graph of $f[n]$ is depicted in Fig. 2.23. Draw the graph of

$$- 2f[2n - 3] \tag{2.7}$$

Solution 2.7 First, let us draw the signal $f[n - 3]$ by shifting the graph of $f[n]$ to the right by 3 units on the time axis as illustrated in Fig. 2.24.

Next, by dividing the time axis of $f[n - 3]$ by 2 and omitting the non-integer division results, we obtain the graph of $f[2n - 3]$ as shown in Fig. 2.25.

In the last step, multiplying the amplitudes of $f[2n - 3]$ by -2, we obtain the graph of $-2f[2n - 3]$ as shown in Fig. 2.26.

It is also possible to write the result we obtained in the form of an integer sequence. We can write the result as an integer sequence as in

Fig. 2.23 The graph of $f[n]$

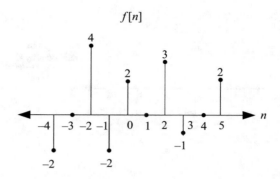

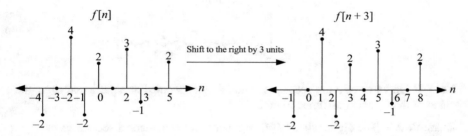

Fig. 2.24 Shifting the digital signal $f[n]$

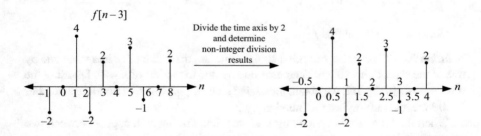

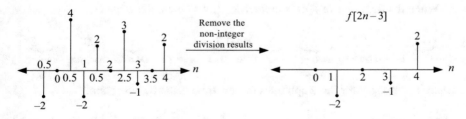

Fig. 2.25 Time scaling on $f[n-3]$

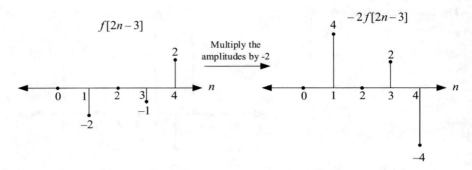

Fig. 2.26 Amplitude scaling on $f[2n - 3]$

$$- 2f[2n - 3] = \left[\underbrace{0}_{n=0} \quad \underbrace{4}_{n=1} \quad \underbrace{0}_{n=2} \quad \underbrace{2}_{n=3} \quad \underbrace{-4}_{n=4} \right]$$

Note If the value of a discrete-time signal is not specified at a particular time, the value of the signal is assumed to be 0.

Example 2.8 The digital signal $f[n]$ is given as a mathematical sequences as

$$f[n] = \left[\underbrace{-1}_{n=-2} \quad \underbrace{2}_{n=-1} \quad \underbrace{-1.5}_{n=0} \quad \underbrace{3}_{n=1} \quad \underbrace{0.5}_{n=2} \quad \underbrace{0}_{n=3} \quad \underbrace{1}_{n=4} \quad \underbrace{0}_{n=5} \quad \underbrace{-2}_{n=6} \right] \quad (2.8)$$

Obtain the digital signal $f\left[\frac{n}{2}\right]$.

Solution 2.8 Time index of digital signal $f[n]$, i.e., the value of n, increases one by one to the right of $n = 0$ and decreases one by one to the left of $n = 0$. Dividing the time axis of $f[n]$ by 1/2, we obtain the digital signal $f\left[\frac{n}{2}\right]$. Note that dividing a number by 1/2 means multiplying that number by 2.

When the time axis is divided by 1/2, new time instants will appear between two elements of $f[n]$. Zero values are assigned to the signal amplitudes for the new time instants.

When the time axis of $f[n]$ is divided by $\frac{1}{2}$, we obtain the sequence

$$\left[\underbrace{-1}_{n=-2 \times 2} \quad \underbrace{2}_{n=-1 \times 2} \quad \underbrace{-1.5}_{n=0 \times 2} \quad \underbrace{3}_{n=1 \times 2} \quad \underbrace{0.5}_{n=2 \times 2} \quad \underbrace{0}_{n=3 \times 2} \quad \underbrace{1}_{n=4 \times 2} \quad \underbrace{0}_{n=5 \times 2} \quad \underbrace{-2}_{n=6 \times 2} \right]$$

where assigning 0 for the amplitudes of new time instants, we obtain.

$$f\left[\frac{n}{2}\right] = \left[\underbrace{-1}_{n=-4} \underbrace{0}_{n=-3} \underbrace{2}_{n=-2} \underbrace{0}_{n=-1} \underbrace{-1.5}_{n=0} \underbrace{0}_{n=1} \underbrace{3}_{n=2} \underbrace{0}_{n=3} \underbrace{0.5}_{n=4} \underbrace{0}_{n=5} \underbrace{0}_{n=6} \underbrace{1}_{n=7} \underbrace{0}_{n=8} \underbrace{0}_{n=9} \underbrace{0}_{n=10} \underbrace{-2}_{n=12} \right]$$

Exercise Using the digital signal $f[n]$ given in the previous example, obtain $f[2n]$.

Exercise Using the digital signal $f[n]$ given in the previous example, we obtain $f\left[\frac{n}{L}\right], L \in$. Write $f\left[\frac{n}{L}\right]$ as a mathematical sequence.

2.4 Sampling

The sampling of continuous-time signals constitutes the basis of digital communication. Sampling is the process of taking samples from a continuous-time signal at certain fixed time intervals. Instead of transmitting the continuous-time signal, we can transmit the samples taken from the continuous-time signal assuming that the sufficient number of samples is taken from the continuous-time signal such that the sample array conveys sufficient information about the continuous-time signal.

We need to collect enough samples. Sending too many samples also means wasting time, wasting energy, and running electronic circuits more than needed.

In this section, we will provide brief information about the sampling operation. Let $f(t)$, $t \in R$ be a continuous-time signal. Taking samples from $f(t)$ at $n \in Z$, $T_s \in R$, nT_s time instants, let us form the mathematical sequence $f[n]$ which is a digital signal. The sampling operation can be explained by the mathematical expression

$$f[n] = f(nT_s), \quad n \in Z, \quad T_s \in R \tag{2.9}$$

where T_s is the sampling period, and its inverse $F_s = \frac{1}{T_s}$ is called sampling frequency.

Sampling of the continuous-time signal $f(t)$ is illustrated in Fig. 2.27. For the sampling operation, the sampling period is chosen as $T_s = 0.5$ ms. As it is seen from Fig. 2.27 that samples are taken from continuous-time signal at time instants which are multiples of T_s, i.e., samples are taken according to, $f[n] = f(nT_S)$, $n \in Z$, $T_s \in R$. Using the taken samples, the mathematical sequence $f[n]$ is formed.

The digital signal, i.e., mathematical sequence, $f[n]$ obtained by sampling operation can be written as

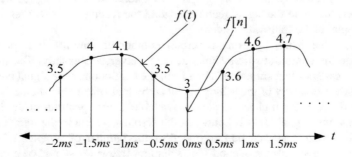

Fig. 2.27 Sampling of a continuous-time signal

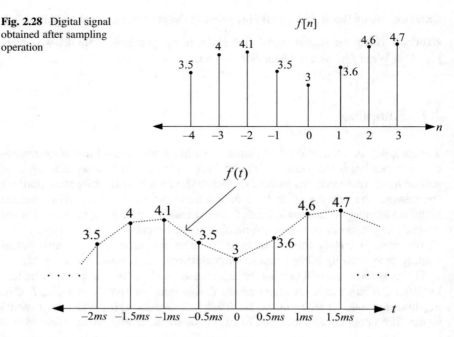

Fig. 2.28 Digital signal obtained after sampling operation

Fig. 2.29 Reconstruction of the analog signal from digital samples

$$f[n] = \left[\begin{array}{cccccccccc} \cdots & \underbrace{3.5}_{n=-4} & \underbrace{4}_{n=-3} & \underbrace{4.1}_{n=-2} & \underbrace{3.5}_{n=-1} & \underbrace{3}_{n=0} & \underbrace{3.6}_{n=1} & \underbrace{4.6}_{n=2} & \underbrace{4.7}_{n=3} & \cdots \end{array} \right]$$

which contains infinitely many numbers.

If we do the sampling operation for a certain time interval, there will be a finite number of elements in our sequence. The digital signal $f[n]$ obtained in Fig. 2.27 can be drawn w.r.t its time index n as in Fig. 2.28.

The samples are sent by the transmitter. On the receiver side, the samples got are used to generate the continuous-time signal sent. This operation is illustrated in Fig. 2.29. As can be seen from the graph in Fig. 2.29, the more samples sent by the transmitter, the more the signal reconstructed at the receiver side will resemble the analog signal at the transmitter side.

On the other hand, by using a certain number of samples, it is possible to re-generate the continuous-time signal at the receiver side exactly. The minimum number of samples that should be taken from the continuous-time signal per second for the perfect recovery of the analog signal at the transmitter side should be greater than two times of the highest frequency available in the spectrum of the low-pass continuous-time signal. This is known as the Nyquist criteria in the literature.

Sampling frequency can be interpreted as the number of samples taken from continuous-time signals per second. The sampling frequency should be greater than two times of the highest frequency available in the spectrum of the low-pass signal.

For example, if the sampling frequency is $F_s = 1000$ Hz, it means that we get 1000 samples from a continuous-time signal in 1 second. Since we take a thousand samples in 1 second, the time interval between two samples is $T_s = \frac{1}{1000}$ sec, which is equal to 1 ms.

Note that $\frac{1}{Hz} = $ sec . We must take a sufficient number of samples from the continuous-time signal in order to make the communication at the lowest satisfactory level. To determine the number of samples to be taken from continuous-time signal, we can use the Nyquist criteria. Let us try to explain the Nyquist sampling criterion with an example.

Example 2.9 Assume that the continuous-time signal

$$f(t) = \sin(200\pi t) + \cos(850\pi t) \tag{2.10}$$

is to be sampled. Determine the minimum sampling frequency such that the perfect reconstruction of the continuous-time signal is possible from its samples.

Solution 2.9 Let us determine the largest frequency within $f(t)$ and choose the sampling frequency as twice the largest frequency. The frequencies of the sinusoidal signals in $f(t)$ can be found by comparing these signals to the mathematical expressions $\sin(2\pi f t)$ and $\cos(2\pi f t)$.

In this case, the frequencies in $f(t)$ are found as $\frac{200}{2} = 100$ Hz and $\frac{850}{2} = 425$ Hz. So, the constraint for the sampling frequency can be written as

$$F_s > 2 \times 425 \rightarrow F_s > 850$$

by which we can determine the sampling frequency as 851. That is, during the digital transmission, at least 851 samples must be sent in 1 second. If we send less than 851 samples per second, the reconstructed signal at the receiver will be different from the $f(t)$.

2.5 Basic Signals Used in Communication and Signal Processing

2.5.1 Representation of Signals with Mathematical Functions

The signal takes values that change over time, and if we draw these values over time, we get a graph of the change of the signal over time. The time changes of the signals can be expressed with mathematical functions. New functions can be obtained by the manipulation of mathematical functions. In other words, functions of functions can be defined, that is, new signals can be generated by processing other signals. In this section, we explain basic continuous and digital signals and see how to create more complex signals from these basic signals.

2.5.2   *Continuous-Time Basic Signal Functions*

Unit Step Function

The unit step function (signal) can be defined in two different ways as in

$$u(t) = \begin{cases} 1 & t > 0 \\ 0 & t < 0 \end{cases} \qquad u(t) = \begin{cases} 1 & t > 0 \\ 0.5 & t = 0.5 \\ 0 & t < 0 \end{cases} \tag{2.11}$$

where for the first definition the value of $u(t)$ for $t = 0$ is not given. The advantage of the second definition is that we can write the equality

$$u(t) + u(-t) = 1$$

which cannot be written using the first definition in (2.11). The graphs of the unit step function for both definitions are given in Fig. 2.30.

Unit Delta Step Function

The unit delta step function (signal) is defined as in

$$u_\Delta(t) = \begin{cases} \dfrac{t}{\Delta} & 0 \le t \le \Delta \\ 1 & t > 0 \\ 0 & \text{otherwise} \end{cases} \tag{2.12}$$

whose graph is depicted in Fig. 2.31.

Fig. 2.30 Unit step function (signal)

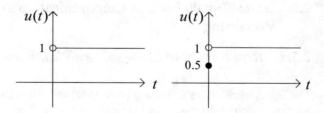

Fig. 2.31   Unit delta step function

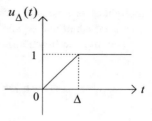

Fig. 2.32   Unit delta impulse function

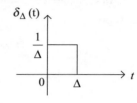

Fig. 2.33   Unit impulse function

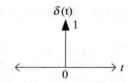

Unit Delta Impulse Function

The unit delta impulse function (signal) is obtained by taking the derivation of the unit delta step function as in

$$\delta_\Delta(t) = \frac{du_\Delta(t)}{dt} \tag{2.13}$$

whose graph is shown in Fig. 2.32.

Unit Impulse Function

Unit impulse function (signal) is obtained from unit delta impulse function as delta goes to zero as illustrated in

$$\delta(t) = \lim_{\Delta \to 0} \delta_\Delta(t) \tag{2.14}$$

The unit impulse function can also be called the impulse function, and its graph is shown in Fig. 2.33. The amplitude of the impulse function equals "1" which indicates the total area under the impulse function. The impulse function can be thought of as a rectangle whose longitude is much longer than the latitude but whose area is constant and equals "1."

The impulse function is mathematically expressed as

$$\delta(t - t_0) = \begin{cases} \infty & t = t_0 \\ 0 & \text{else} \end{cases} \quad \text{with} \quad \int \delta(t - t_0) dt = 1 \tag{2.15}$$

An equivalent definition can be given as

$$\delta(t) = \begin{cases} \infty & t = t_0 \\ 0 & \text{else} \end{cases} \quad \text{with} \quad \int \delta(t) dt = 1 \tag{2.16}$$

2.5.3 Properties of the Impulse Function

The impulse function is one of the basic functions widely used in communication and signal processing. In this respect, some basic properties of the impulse function should be well understood.

Properties of the impulse function can be listed as follows:

1. $\delta(t - t_0)f(t) = \delta(t - t_0)f(t_0)$, where t_0 is a real number and t represents the time parameter.
2. $\delta(t - t_0)f(t - t_1) = \delta(t - t_0)f(t_0 - t_1)$ is the more general form of (1).
3. $\int_{-\infty}^{\infty} \delta(t) dt = 1$ or $\int_{-\infty}^{\infty} \delta(t - t_0) dt = 1$.
4. $\delta(at) = \frac{1}{|a|} \delta(t) \quad \delta(a(t - t_0)) = \frac{1}{|a|} \delta(t - t_0)$
5. $\int_{-\infty}^{\infty} \delta(t - t_0)f(t) dt = f(t_0)$ or $\int_{-\infty}^{\infty} \delta(t - t_0)f(t - t_1) dt = f(t_0 - t_1)$.

The proof of (5) can be achieved using (1), (2), and (3). Now let us prove (5) as an exercise.

From (1), we can write

$$\int_{-\infty}^{\infty} \delta(t - t_0)f(t) dt = \int_{-\infty}^{\infty} \delta(t - t_0)f(t_0) dt \rightarrow$$

$$\int_{-\infty}^{\infty} \delta(t - t_0)f(t_0) dt = f(t_0) \underbrace{\int_{-\infty}^{\infty} \delta(t - t_0) dt}_{=1}$$

and from (3) we have

$$f(t_0) \int\limits_{-\infty}^{\infty} \delta(t - t_0)dt = f(t_0)$$

1. , 2. , and n. derivatives of the impulse function are shown as $\dot{\delta}(t)$, $\ddot{\delta}(t)$, and $\delta^n(t)$. Using the n. derivative of the impulse function, we can write the properties

$$f(t)\delta^n(t - t_0) = \frac{\partial^n f(t)}{\partial t^n}\bigg|_{t=t_0} \delta(t - t_0) \tag{2.17}$$

$$\int\limits_{t=-\infty}^{\infty} f(t)\delta^n(t - t_0)dt = \frac{\partial^n f(t)}{\partial t^n}\bigg|_{t=t_0} \tag{2.18}$$

Ramp Function

Ramp function is defined as

$$r(t) = \begin{cases} t & t \geq 0 \\ 0 & \text{else} \end{cases} \tag{2.19}$$

The graph of the ramp function is shown in Fig. 2.34.

The mathematical relationships between unit step function, impulse function, and ramp function can be expressed as

$$u(t) = \frac{dr(t)}{dt} \quad r(t) = \int\limits_{-\infty}^{t} u(t)dt \quad \delta(t) = \frac{du(t)}{dt}$$

$$\tag{2.20}$$

$$u(t) = \int\limits_{-\infty}^{t} \delta(t)dt \quad \delta(t) = \frac{\partial^2 r(t)}{\partial^2 t}$$

Now let us give some examples to reinforce these basic functions we have learned.

Fig. 2.34 Ramp function

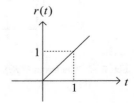

Example 2.10 Using $f(t) = 2t^2 + 1$, find the results of the calculations given below:

(a) $f(t)\delta(t - 1) = ?$
(b) $f(t)\dot{\delta}(t - 2) = ?$
(c) $\int_{t=-\infty}^{\infty} f(t)\delta(t)dt = ?$
(d) $\int_{t=-\infty}^{\infty} f(t)\delta(t - 2)dt = ?$
(e) $\int_{t=-\infty}^{\infty} f(t)\dot{\delta}(t - 1)dt = ?$

Solution 2.10
(a) $f(t)\delta(t - 1) = f(1)\delta(t - 1) \rightarrow (2.1^2 + 1)\delta(t - 1) = 3\delta(t - 1)$
(b) $f(t)\dot{\delta}(t - 2) = \frac{df(t)}{dt}\Big|_{t=2}\delta(t - 2) \rightarrow \left(4t|_{t=2}\right)\delta(t - 2) = 8\delta(t - 2)$
(c) $\int_{t=-\infty}^{\infty} f(t)\delta(t)dt = f(0) \rightarrow 2.0^2 + 1 = 1$
(d) $\int_{t=-\infty}^{\infty} f(t)\delta(t - 2)dt = f(2) \rightarrow 2.2^2 + 1 = 9$
(e) $\int_{t=-\infty}^{\infty} f(t)\dot{\delta}(t - 1)dt = \frac{df(t)}{dt}\Big|_{t=1} \rightarrow \left(4t|_{t=1}\right) = 4$

Example 2.11 Find the results of the operations given below:

(a) $\frac{du(t-2)}{dt} = ?$

(b) $\frac{du(t^2+1)}{dt} = ?$

(c) $\frac{dr(t^2+1)}{dt} = ?$

Solution 2.11
(a) $\frac{du(t-2)}{dt} = \delta(t - 2)$
(b) $\frac{du(t^2+1)}{dt} = \frac{d(t^2+1)}{dt}\delta(t^2 + 1) \rightarrow (2t)\delta(t^2 + 1)$
(c) $\frac{dr(t^2+1)}{dt} = \frac{d(t^2+1)}{dt}u(t^2 + 1) \rightarrow (2t)u(t^2 + 1)$

Example 2.12 Draw the graph of $f(t) = u(t - 1) - u(t - 3) + \delta(t - 2)$.

Solution 2.12 To draw the graph of $f(t) = u(t - 1) - u(t - 3) + \delta(t - 2)$, we first draw the graph of $u(t - 1) - u(t - 3)$, and then adding the graph of the shifted impulse $\delta(t - 2)$, we get the graph of $f(t)$ as in Fig. 2.35.

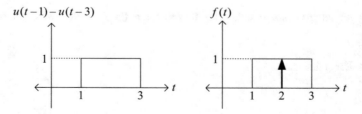

Fig. 2.35 The graph of $f(t) = u(t - 1) - u(t - 3) + \delta(t - 2)$

Fig. 2.36 A ramp-like
signal

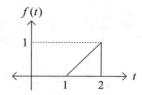

Fig. 2.37 Shifted and
negative amplitude scaled
ramp signals

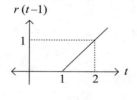

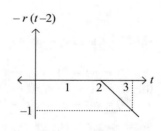

Fig. 2.38 The graph of
$r(t - 1) - r(t - 2)$

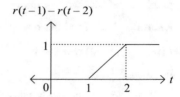

Example 2.13 Express the function whose graph is given in Fig. 2.36 in terms of the ramp and unit step functions.

Solution 2.13 The graphs of $r(t - 1)$ and $-r(t - 2)$ are shown in Fig. 2.37.

The slopes of the shifted ramp functions depicted in Fig. 2.37 are 1 and -1, respectively. If the functions shown in Fig. 2.37 are summed, we obtain the function shown in Fig. 2.38 where it is seen that after $t = 2$, we have a horizontal line with slope 0.

It is seen from Fig. 2.38 that from the time instant $t = 2$ on, the amplitude of the function value is 1, whereas, in $f(t)$ depicted in Fig. 2.36, the amplitude value is 0 from the time instant $t = 2$ on.

In order to make the amplitude 0 from the time instant $t = 2$ on, it is necessary to add function $-u(t - 2)$ to the function in Fig. 2.38. This operation is explained in Fig. 2.39.

Example 2.14 Draw the graphs of the functions given below:

(a) $f(t) = \delta(t + 2) + \delta(t - 3)$
(b) $f(t) = -\delta(t + 3) + \delta(t)$
(c) $f(t) = u(t - 3)$
(d) $f(t) = u(t - 1) + u(t - 4)$
(e) $f(t) = \delta(t + 1) + u(t - 1)$
(f) $f(t) = \delta(t - 2) + u(t - 1)$

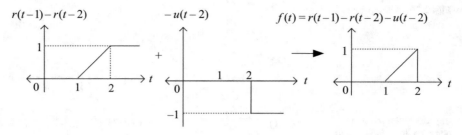

Fig. 2.39 Obtaining graph of $r(t-1) - r(t-2) - u(t-2)$

(g) $f(t) = r(t-3)$
(h) $f(t) = r(t-1) + r(t-2) - 2r(t-5)$
(i) $f(t) = \delta(t-3) + r(t-1)$

Solution 2.14 The graphs of the functions are drawn in Fig. 2.40. Inspecting the graphs carefully, we can understand how they are drawn.

Example 2.15 Find the results of the integrals in (a) and (b):

(a) $\int_{-\infty}^{t} \delta(t-2)dt = ?$
(b) $\int_{-\infty}^{t} u(t-2)dt = ?$

Write an explicit mathematical expression for the function $f(t) = \delta(t^2 - 4)$.

Solution 2.15
(a) $\int_{-\infty}^{t} \delta(t-2)dt = u(t-2)$
(b) $\int_{-\infty}^{t} u(t-2)dt = r(t-2)$
(c) $f(t) = \delta(t^2 - 4) = \begin{cases} \infty & t = -2 \text{ or } t = 2 \\ 0 & \text{else} \end{cases}$

Example 2.16 Express the functions whose graphs are shown in Fig. 2.41 in terms of the unit step functions.

Solution 2.16 Functions can be easily written in terms of unit step functions by considering the points where discontinuities occur in the graphs and by calculating the amount of discontinuity at the discontinuity points:

(a) $f(t) = 2u(t-2) + 2u(t-5)$
(b) $f(t) = 2u(t-1) + 2u(t-3) - 4u(t-5)$
(c) $f(t) = 2u(t-2) - 2u(t-5)$
(d) $f(t) = 2u(t-2) - 4u(t-4) + 2u(t-6)$
(e) $f(t) = 2u(t+2) + 2u(t-4) - 2u(t-6)$
(f) $f(t) = 4u(t-2) - 2u(t-3) + 2u(t-4) - 4u(t-5)$

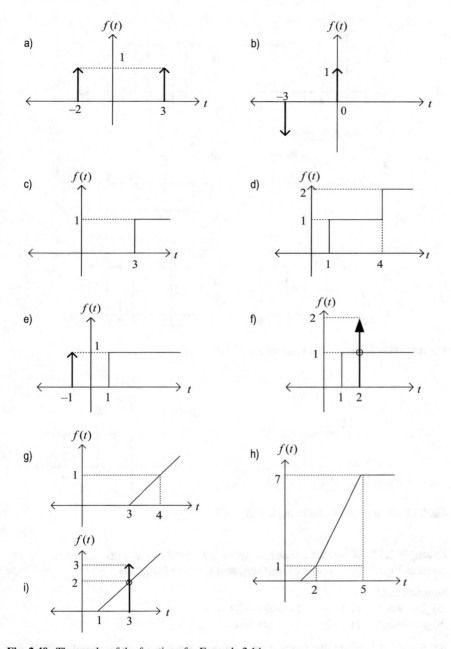

Fig. 2.40 The graphs of the functions for Example 2.14

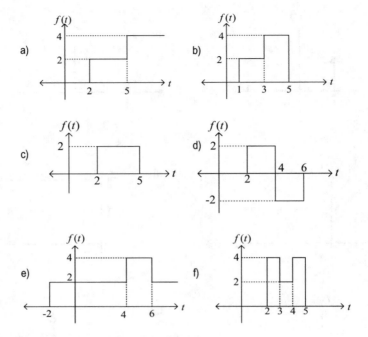

Fig. 2.41 Function graphs for Example 2.16

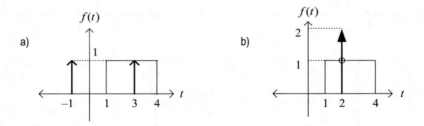

Fig. 2.42 Function graphs for Example 2.17

Example 2.17 Write mathematical expressions for the functions, whose graphs are given in Fig. 2.42, in terms of unit step and impulse functions.

Solution 2.17
(a) $f(t) = \delta(t + 1) + u(t - 1) + \delta(t - 3) - u(t - 4)$
(b) $f(t) = u(t - 1) + \delta(t - 2) - u(t - 4)$

Example 2.18 Write mathematical expressions for the functions, whose graphs are given in Fig. 2.43, in terms of unit step, impulse, and ramp functions.

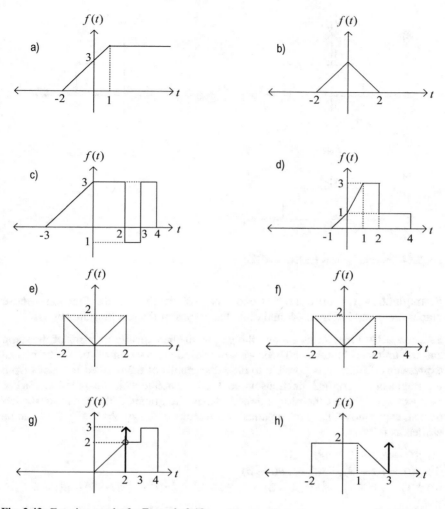

Fig. 2.43 Function graphs for Example 2.18

Solution 2.18 In such questions, first we determine the points on the horizontal axis where the amplitude of the function changes significantly. Then, paying attention to those points, we can express the function in terms of the basic functions as follows:

(a) $f(t) = r(t + 2) - r(t - 1)$
(b) $f(t) = r(t + 2) - 2r(t) + r(t - 2)$
(c) $f(t) = r(t + 3) - r(t) - 4u(t - 2) + 4u(t - 3) - 3u(t - 4)$
(d) $f(t) = r(t + 1) + r(t) - 2r(t - 1) - 2u(t - 2) - u(t - 4)$
(e) $f(t) = r(-t) - r(-t - 2) - 2u(-t - 2) + r(t) - r(t - 2) - 2u(t - 2)$
(f) $f(t) = r(-t) - r(-t - 2) - 2u(-t - 2) + r(t) - r(t - 2) - 2u(t - 4)$
(g) $f(t) = r(t) + \delta(t - 2) - r(t - 2) + u(t - 3) - 3u(t - 4)$
(h) $f(t) = 2u(t + 2) - r(t - 1) - r(t - 3) + \delta(t - 3)$

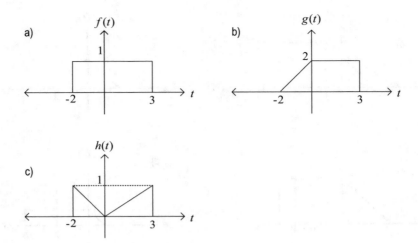

Fig. 2.44 Function graphs for Example 2.19

Example 2.19 Calculate the first and second derivatives of the functions whose graphs are given in Fig. 2.44, and draw the graphs of the resulting functions.

Solution 2.19 First, let us express the graphs mathematically in terms of the ramp and the unit step functions. Then, we can take the derivation of the mathematical expressions. In fact, it is possible to draw the graphs of the derived functions from the graphs of the original functions without writing mathematical expressions. In our next example, we get the results directly using the graphs without writing mathematical expressions. The mathematical expressions of the graphs in Fig. 2.44 can be written as follows:

(a) $f(t) = u(t + 2) - u(t - 3)$
(b) $g(t) = r(t + 2) - r(t) - 2u(t - 3)$
(c) $h(t) = \frac{1}{2} r(-t) - \frac{1}{2} r(-t-2) - u(-t-2) + \frac{1}{3} r(t) - \frac{1}{3} r(t-3) - u(t-3)$

We can take the derivatives of the mathematical functions in the previous sentence as follows:

(a) $\dot{f}(t) = \delta(t + 2) - \delta(t - 3)$
(b) $\dot{g}(t) = u(t + 2) - u(t) - 2\delta(t - 3)$
(c) $\dot{h}(t) = -\frac{1}{2} u(-t) + \frac{1}{2} u(-t-2) + \delta(-t-2) + \frac{1}{3} u(t) - \frac{1}{3} u(t-3) - \delta(t-3)$

The graphs of the derived functions can be drawn as in Fig. 2.45.

If we inspect the derivative graphs, we see that the impulse function appears at time instants '*t*' where the original function, whose derivative is taken, has discontinuities. To plot the graph of the derivative function, we can consider the two drawing rules as follows:

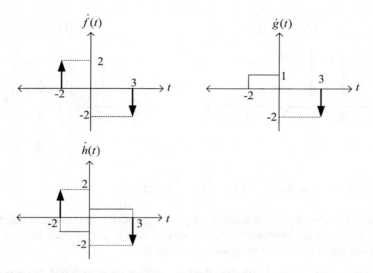

Fig. 2.45 The graphs of the derived functions for Example 2.18

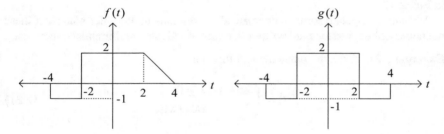

Fig. 2.46 Function graphs for Example 2.20

Rule 1 We determine the discontinuity points, and calculate the amount of amplitude differences at the discontinuity points. If there is an upward jump at the discontinuity point, an impulse function with an amplitude equal to the amount of amplitude difference at the discontinuity point appears in the derivative graph. If the discontinuity is downward, a downward impulse function, whose amplitude equals to the amplitude difference, appears in the derivative graph at the discontinuity point.

Rule 2 If there are sloping lines in the graph to be derived, the slope amounts are calculated and indicated on the graph as horizontal lines.

Example 2.20 Calculate the derivatives of the functions given in Fig. 2.46 directly using the graph without writing a formula.

Solution 2.20 Let us find the derivatives of the graphics directly by applying only rule 1 and rule 2 without writing formulas for the graphics. For this purpose, we should determine the discontinuity points on the graphs and calculate the amount of

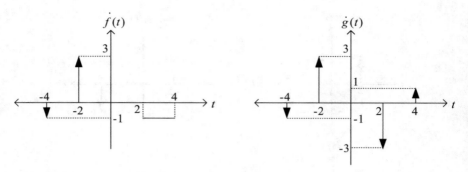

Fig. 2.47 Graphs of the derivative function for Example 2.20

amplitude differences at the discontinuity points, and draw downward impulses with amplitudes equal to downward discontinuity amounts and upward impulses with amplitudes equal to upward discontinuity amounts.

For lines with slopes, horizontal lines are drawn by calculating the amount of slopes. Applying these rules, the graphics of the derivative functions can be drawn as in Fig. 2.47.

We can manipulate some mathematical expressions in an easier manner if these mathematical expressions are written in terms of unit step and impulse functions.

Example 2.21 Write the mathematical function

$$f(t) = \begin{cases} 2t + 1 & t \geq 0 \\ 0 & \text{otherwise} \end{cases} \tag{2.21}$$

in terms of the unit step function.

Solution 2.21 $f(t) = (2t + 1)u(t)$

Example 2.22 Write the mathematical function

$$f(t) = \begin{cases} 2t - 1 & t \geq 0 \\ -2 & t < 0 \\ 0 & \text{otherwise} \end{cases} \tag{2.22}$$

in terms of unit step functions.

Solution 2.22 $f(t) = (2t - 1)u(t) - 2u(-t)$

Example 2.23 Express the functions

$$\delta(2t)\delta(-2t)\delta(2t-1)\delta(3t-3)$$

in terms of $\delta(t)$.

Solution 2.23

$$\delta(2t) = \frac{1}{2}\delta(t) \quad \delta(-2t) = \frac{1}{2}\delta(t)$$

$$\delta(2t-1) = \frac{1}{2}\delta\left(t-\frac{1}{2}\right) \quad \delta(3t-3) = \frac{1}{3}\delta(t-1)$$

Example 2.24 Draw the graph of the function $\delta[2n-3]$.

Solution 2.24 Using the definition of impulse function, we can write the given function as

$$\delta[2n-3] = \begin{cases} 1 & \text{if } 2n-3 = 0 \rightarrow n = \dfrac{3}{2} \\ 0 & \text{otherwise} \end{cases}$$

where n is found as $\frac{3}{2}$ which is not an integer. In this case, the graph of the given function does not exist.

Example 2.25 Draw the graph of $\overset{\circ}{\delta}(t)$, which is the derivative of impulse function, and comment on the drawn graph.

Solution 2.25 As we mentioned earlier, the impulse function can be written as the limit of the unit delta impulse function. That is,

$$\delta(t) = \lim_{\Delta \to 0} \delta_\Delta(t) \tag{2.23}$$

where $\delta_\Delta(t)$ is the delta impulse function whose graph can be drawn as in Fig. 2.48.

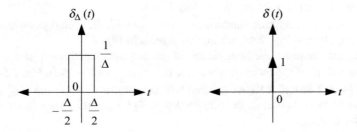

Fig. 2.48 Graphs of $\delta_\Delta(t)$ and $\delta(t)$

Fig. 2.49 Graph of $\dot{\delta}_\Delta(t)$

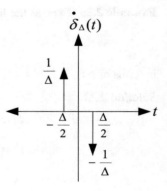

Fig. 2.50 Graph of $\dot{\delta}_\Delta(t)$

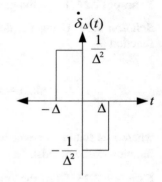

Taking the derivative of both sides of (2.23), we obtain

$$\dot{\delta}(t) = \lim_{\Delta \to 0} \dot{\delta}_\Delta(t)$$

So, in order to plot the function $\dot{\delta}(t)$, let us first plot the function $\dot{\delta}_\Delta(t)$. Using the discontinuity points of the delta impulse function given in Fig. 2.48, the graph of the derivative of the delta impulse function $\dot{\delta}_\Delta(t)$ can be drawn as in Fig. 2.49.

The amplitudes of the impulse functions in Fig. 2.49 are $\frac{1}{\Delta}$ and $-\frac{1}{\Delta}$. That is, the areas under the impulse functions are $\frac{1}{\Delta}$ and $-\frac{1}{\Delta}$.

If we replace the impulse functions in Fig. 2.49 with the rectangles whose limits give the impulses in Fig. 2.49, we get the signals in Fig. 2.50.

Taking into account the area values of the impulse functions $\frac{1}{\Delta}$ and $-\frac{1}{\Delta}$, the amplitude values of the rectangles are calculated as $\frac{1}{\Delta^2}$ and $-\frac{1}{\Delta^2}$ in Fig. 2.50. If we compare Figs. 2.48 and 2.50, we see that both amplitudes of $\dot{\delta}_\Delta(t)$ and $\delta_\Delta(t)$ go to ∞ as Δ goes to zero; however, the amplitudes of $\dot{\delta}_\Delta(t)$ go to infinity faster than the amplitudes of $\delta_\Delta(t)$.

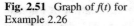

Fig. 2.51 Graph of $f(t)$ for Example 2.26

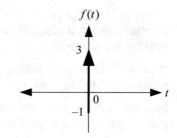

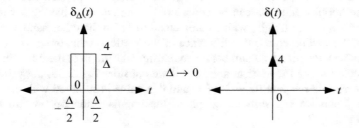

Fig. 2.52 Graphs of $\delta_{\Delta}(t)$ and $\delta(t)$

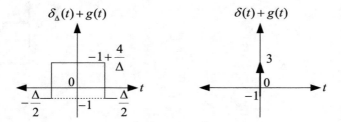

Fig. 2.53 Graphs of $\delta_{\Delta}(t) + g(t)$ and $\delta(t) + g(t)$

Exercise Draw the graph of $\ddot{\delta}(t)$ and compare it to the graph of $\dot{\delta}(t)$. Comment on $\ddot{\delta}(t)$ and compare it to $\dot{\delta}(t)$.

Example 2.26 The graph of the impulse function shifted on the vertical axis is given in Fig. 2.51. Comment on this function.

Solution 2.26 The area of the impulse function given in Fig. 2.51 is 4. In Fig. 2.52 the graph of an un-shifted impulse function with an area of 4 is depicted.

Let us define $g(t)$ function as

$$g(t) = \begin{cases} -1 & \text{if } -\dfrac{\Delta}{2} \le t \le \dfrac{\Delta}{2} \\ 0 & \text{otherwise} \end{cases} \tag{2.24}$$

In Fig. 2.53, the graph of the function resulting from the summation of impulse function and $g(t)$ is depicted. It is seen from Fig. 2.53 that, as $\delta \rightarrow 0$, we obtain the impulse function shifted along the vertical axis.

Exercise The graph of the impulse function shifted along the vertical and horizontal axes is given in Fig. 2.54. Comment on this function.

Example 2.27 In Fig. 2.55, the summation of the impulse function and the rectangular function is depicted. Comment on this graph.

Solution 2.27 The amplitude of $r(t)$ is 1. The area under the impulse function is 1 as well. The impulse function can be considered as a rectangle, having a very small width and very large height, with an area equal to 1. On the other hand, the area of the impulse function equals the difference of the vertical ordinates.

Summing a function by a number means shifting the graph of the function upward or downward along the vertical axis. So, if we consider Fig. 2.55, the value of the rectangle at time $t = 2$ is $r(t = 2) = 1$, and this value is summed with the impulse function. Then we can shift the graph of the impulse function upward by 1 as depicted in Fig. 2.56.

Fig. 2.54 Function graph for exercise

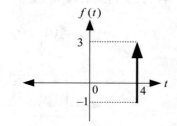

Fig. 2.55 Graph of a function for Example 2.27

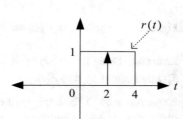

Fig. 2.56 Upward shifted impulse

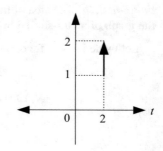

Fig. 2.57 Graph of $r(t)$ with a missing point

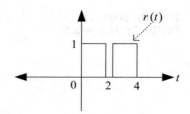

Fig. 2.58 Graph of $r(t)$ with a missing point

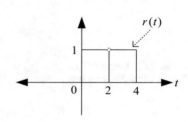

Fig. 2.59 Graph of summed functions

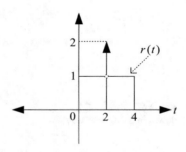

We used the rectangle's value at $t = 2$, $r(t = 2) = 1$, to shift the impulse function up. So if we subtract this point from the rectangle function, the rectangle will look as if it is divided into two parts, that is, the rectangle becomes as in Fig. 2.57.

where we have $r(2^-) = 1$, $r(2^+) = 1$ and $r(2) = 0$. The graphic in Fig. 2.57 is exaggerated. It can be drawn as in Fig. 2.58 where a single missing point is indicated by a circle.

Finally, if we sum up the graph of the shifted impulse function and the graph in Fig. 2.58, we obtain Fig. 2.59.

Example 2.28 The graph of $f(t)$ is depicted in Fig. 2.60. Draw the graph of $g(t) = f(t) + f(t - 2)$.

Solution 2.28 Looking at the graph of the function $f(t)$, it is seen that the values of the function at $t = -1$ and $t = 1$ are $f(-1) = -1$ and $f(1) = 2$. Vertical lines are used at the time instants $t = -1$ and $t = 1$ where the function gets values 1 and 2. It is possible to draw the function in Fig. 2.60 as in Fig. 2.61.

In Fig. 2.61, we specifically highlighted the values at the extreme points with bold black dots. Fig. 2.61 can also be drawn as in Fig. 2.62 without using black dots at the end of the lines.

Fig. 2.60 Graph of a
function for Example 2.28

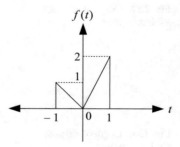

Fig. 2.61 Graph of f(t)
drawn in a different way

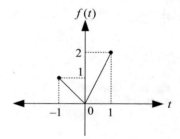

Fig. 2.62 Graph of f(t)
drawn in a different way

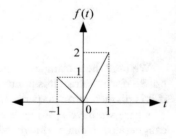

If the function does not have value at $t = -1$ and $t = 1$, then its graph is drawn as in Fig. 2.63.

Let us now draw the graph of $g(t) = f(t) + f(t - 2)$. To draw the graph of the function $f(t - 2)$, it is sufficient to shift the values of the function $f(t)$ by 2 units to the right along the time axis. The graphs of $f(t)$ and $f(t - 2)$ are shown in Fig. 2.64.

It is seen from Fig. 2.64 that the functions $f(t)$ and $f(t + 2)$ overlap only at the point $t = 1$. There are no other overlapping points. Thus, the value of $g(t)$ at point $t = 1$ can be calculated as $g(1) = 1 + 2$. Accordingly, the graph of $g(t)$ can be drawn as in Fig. 2.65.

The graph in Fig. 2.65 can also be drawn as in Fig. 2.66.

Example 2.29 The graph of $f(t)$ is shown in Fig. 2.67. Draw the graph of $g(t) = f(t - 1) + f(t) + f(t + 1)$.

Fig. 2.63 Graph of $f(t)$ having missing points

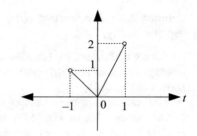

Fig. 2.64 Graphs of $f(t)$ and $f(t - 2)$

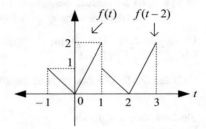

Fig. 2.65 Graph of $g(t)$

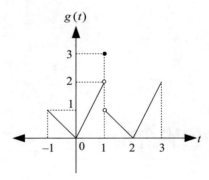

Fig. 2.66 Graph of $g(t)$ using alternative drawing

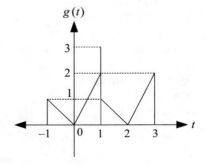

Solution 2.29 The function $g(t)$ contains $f(t-1)$, $f(t)$ and $f(t+1)$ whose graphs are depicted in Fig. 2.68.

For the addition of the functions shown in Fig. 2.68, we first need to determine the overlapping lines and write the equations of these lines. The overlapping lines are shown in Fig. 2.69.

The equations of the overlapping lines are indicated in Fig. 2.70.

On the left side of Fig. 2.70, the equations of the overlapping lines are $t+1$ and $-t$ in the interval they are defined, and adding these two equations, we get $t+1-t=1$. Similarly, if the equations of the two lines that overlap on the right side of Fig. 2.70 are added, we get $t+(-t+1)=1$.

Thus, we can draw the graph of $g(t)$ as in Fig. 2.71.

Example 2.30 The graph of $f(t)$ is shown in Fig. 2.72. Draw the graph of $g(t)=\sum_{k=-\infty}^{\infty} f(t-k)$.

Solution 2.30 We can expand the summation

Fig. 2.67 Graph of $f(t)$ for Example 2.29

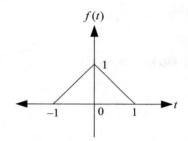

Fig. 2.68 Graphs of $f(t-1)$, $f(t)$ and $f(t+1)$

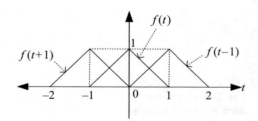

Fig. 2.69 The overlapping lines

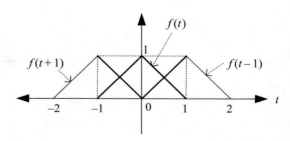

Fig. 2.70 Equations for the overlapping lines

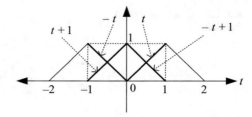

Fig. 2.71 Graph of $f(t-1) + f(t) + f(t+1)$

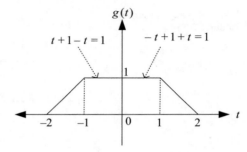

Fig. 2.72 Graph of $f(t)$ for Example 2.30

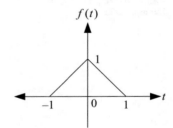

$$g(t) = \sum_{k=-\infty}^{\infty} f(t-k)$$

as

$$g(t) = \cdots + f(t+2) + f(t+1) + f(t) + f(t-1) + f(t-2) + \cdots$$

where the graphs of the shifted functions can be drawn as in Fig. 2.73.

Summing the equations of the overlapping lines in Fig. 2.73, we obtain the graph of $g(t)$ as in Fig. 2.74.

Example 2.31 The graph of $f(t)$ is shown in Fig. 2.75. Draw the graph of $g(t) = f(t) + f(t-1.5)$.

Solution 2.31 We can draw the graph of

Fig. 2.73 Graphs of shifted functions

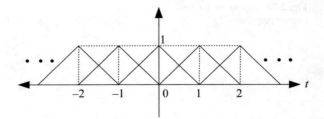

Fig. 2.74 Graph of $g(t)$ for Example 2.30

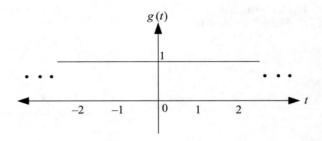

Fig. 2.75 Graph of $f(t)$ for Example 2.31

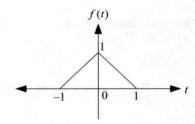

Fig. 2.76 Graphs of $f(t)$ and $f(t-1.5)$

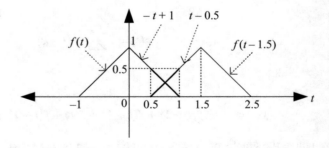

$$g(t) = f(t) + f(t - 1.5)$$

as illustrated in Figs. 2.76 and 2.77.

Fig. 2.77 Graph of
$f(t) + f(t - 1.5)$

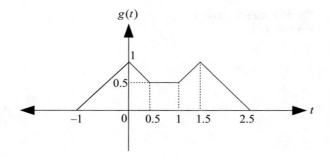

Fig. 2.78 Graph of $f(t)$ for
exercise

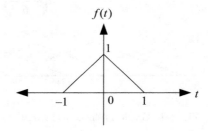

Exercise The graph of $f(t)$ is depicted in Fig. 2.78. Draw the graph of

$$g(t) = \sum_{k=-\infty}^{\infty} f(t - 1.5k).$$

Exercise Draw the graph of

$$g(t) = \sum_{k=-\infty}^{\infty} f(t - 1.3k)$$

Using $f(t)$ in Fig. 2.78.

Example 2.32 The graph of $X(w)$ is given in Fig. 2.79. Draw the graph of

$$Y(w) = \sum_{k=-\infty}^{\infty} X\left(\frac{w - k2\pi}{6}\right).$$

Solution 2.32 If we expand the summation

$$Y(w) = \sum_{k=-\infty}^{\infty} X\left(\frac{w - k2\pi}{6}\right)$$

giving values to k, we obtain

Fig. 2.79 Graph of $X(w)$ for Example 2.32

Fig. 2.80 Graphs of shifted and scaled $X(w)$ signals

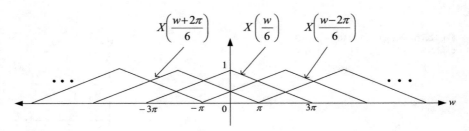

Fig. 2.81 Repeating pattern

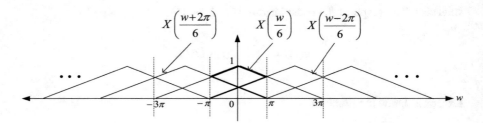

$$Y(w) = \cdots + X\left(\frac{w+2\pi}{6}\right) + X\left(\frac{w}{6}\right) + X\left(\frac{w-2\pi}{6}\right) + \cdots$$

First, let us draw the graph of $X\left(\frac{w}{6}\right)$. Shifting the graph of $X\left(\frac{w}{6}\right)$ to the right by 2π, we obtain the graph of $X\left(\frac{w-2\pi}{6}\right)$, and shifting the graph of $X\left(\frac{w}{6}\right)$ to the left by 2π, we get the graph of $X\left(\frac{w+2\pi}{6}\right)$.

Accordingly, the graph of $Y(w)$ can be obtained by summing all the shifted graphs and the central one as depicted in Fig. 2.80.

The summation of the graphs in Fig. 2.80 may seem difficult at first glance. However, when we inspect Fig. 2.80, we see that a certain unit is repeated throughout the shape. The repeating unit is shown in bold in Fig. 2.81.

We first sum the equations of the crossing lines in Fig. 2.81. Since the slopes of these lines are of equal and opposite signs, the result of the sum will be a fixed number. In Fig. 2.81, around the center, the equation of the diagonal line with a positive slope is

$$\frac{w}{3\pi} + \frac{1}{3}$$

and the line with a negative slope has the equation

$$-\frac{w}{3\pi} + \frac{1}{3}$$

The sum of the equations is

$$\frac{w}{3\pi} + \frac{1}{3} - \frac{w}{3\pi} + \frac{1}{3} = \frac{2}{3}.$$

In other words, when the equations of the crossing lines around the center are summed, we get the graph in Fig. 2.82.

Finally, when the triangle and the horizontal line equations in Fig. 2.82 are summed, we obtain Fig. 2.83.

To obtain the graph of $Y(w)$, we shift the graph in Fig. 2.83 to the right and left by multiples of 2π and sum the shifted replicas with the centered graph. The result of the summation is depicted in Fig. 2.84.

Fig. 2.82 Simplified repeating pattern

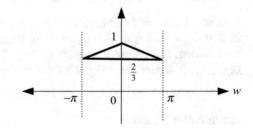

Fig. 2.83 Final form of the repeating pattern

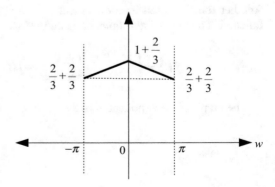

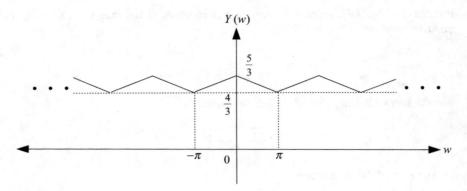

Fig. 2.84 Graph of $Y(w)$ for Example 2.32

Exercise Draw the graphs of the functions given as

(a) $u(2t)$
(b) $u(3t)$
(c) $u(-t)$
(d) $\delta(-t)$
(e) $\delta(-2t)$
(f) $\delta(-3t)$
(g) $\delta(-2t+1)$

Example 2.33 Show that $\delta(at-b)=\frac{1}{|a|}\delta\left(t-\frac{b}{a}\right)$.

Solution 2.33 The function $\delta(at-b)$ is obtained from $\delta_\Delta(at-b)$ as $\Delta\to 0$. So let us first draw the expression $\delta_\Delta(at-b)$, then calculate the area of this function we have drawn, and consider its limit as $\Delta\to 0$. What we are describing is shown in Fig. 2.85. It is seen from Fig. 2.85 that the area of $\delta(at-b)$ equals $\frac{1}{|a|}$.

Exponential Function

Another function used in communication and signal processing is the exponential function. The exponential function is defined as

$$f(t)=\begin{cases} ke^{-t} & t\geq 0 \\ 0 & \text{otherwise} \end{cases} \rightarrow f(t)=ke^{-t}u(t), k\in R \qquad (2.25)$$

The graph of the exponential function for $k=1$, $0\leq t\leq 2$ is shown in Fig. 2.86.

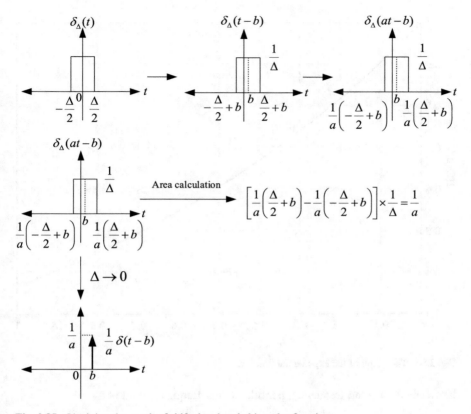

Fig. 2.85 Obtaining the graph of shifted and scaled impulse function

Sinusoidal Function

Continuous-time sinusoidal function is defined as

$$f(t) = K \cos\left(\frac{2\pi}{T}t + \theta\right) \tag{2.26}$$

where T is the period of the function, $f = \frac{1}{T}$ is the frequency of the function (signal), and the parameter θ indicates the phase difference. The angular frequency is defined as $w = \frac{2\pi}{T}$, and using the angular frequency, we can define the continuous-time sinusoidal function as $f(t) = K \cos(wt + \theta)$.

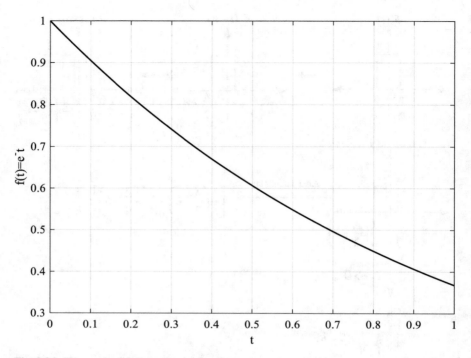

Fig. 2.86 The graph of the exponential function

Example 2.34 Find frequency, period, angular frequency, and phase of

$$f(t) = \cos\left(\frac{\pi}{3}t + \frac{\pi}{2}\right).$$

Solution 2.34 When the given function is compared to $f(t) = K \cos(wt + \theta)$ we see that $w = \frac{\pi}{3} \rightarrow \frac{2\pi}{T} = \frac{\pi}{3}$ from here, we obtain $T = 6$, frequency can be calculated as $f = \frac{1}{T} \rightarrow f = \frac{1}{6}$, the phase difference is $\theta = \frac{\pi}{2}$, and angular frequency can be calculated as $w = \frac{\pi}{3}$. Unless otherwise stated, the unit of the period is the second, and the unit of the frequency is Hertz.

Exponentially Damped Sinusoidal Function

The exponentially damped sinusoidal function is defined as

$$f(t) = Ke^{-\alpha t} \cos\left(\frac{2\pi}{T}t + \theta\right) \tag{2.27}$$

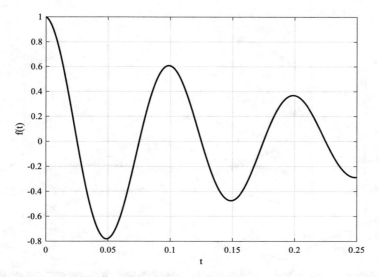

Fig. 2.87 The graph of the exponentially damped function

where the exponential function $e^{-\alpha t}$ is multiplied by the sinusoidal function $K \cos(wt + \theta)$. In Fig. 2.87, the exponentially damped sinusoidal function $K = 1$, $\alpha = 5, \theta = 0, T = 0.1$ is depicted.

Complex Exponential Function

The complex exponential function is defined as

$$f(t) = \begin{cases} ke^{jwt} & t \geq 0 \\ 0 & \text{otherwise} \end{cases} \rightarrow f(t) = ke^{jwt}u(t), k \in R \qquad (2.28)$$

where $e^{jwt} = \cos(wt) + j \sin(wt)$, and using

$$\left| e^{jwt} \right| = \sqrt{\cos^2(wt) + \sin^2(wt)} \rightarrow \left| e^{jwt} \right| = 1$$

we obtain $|f(t)| = k, t \geq 0$. It is possible to draw the graph of the complex exponential function on a three-dimensional plane. In Fig. 2.88, for $w = 8\pi$ and $k = 1$, the complex exponential function is drawn in three-dimensional space.

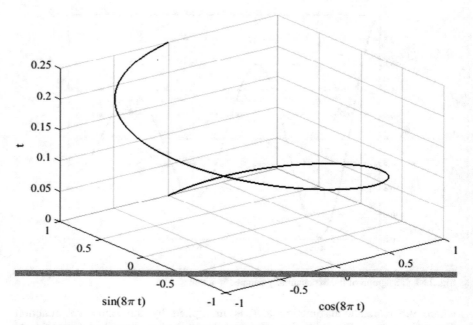

Fig. 2.88 The graph of the complex exponential function

Fig. 2.89 Digital unit step
function

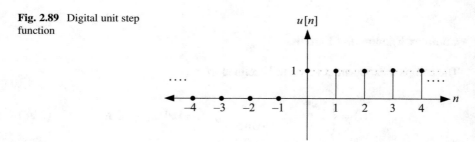

2.5.4   Basic Digital Signal Functions

Unit Step Function

Digital unit step function is defined as

$$u[n] = \begin{cases} 1 & n \geq 0, n \in Z \\ 0 & \text{otherwise} \end{cases} \tag{2.29}$$

where Z denotes the set of integers. The graph of (2.29) is shown in Fig. 2.89.

Unit Impulse Function

Digital unit impulse function is defined as

$$\delta[n - n_0] = \begin{cases} 1 & n = n_0 \\ 0 & \text{otherwise} \end{cases} \tag{2.30}$$

We can also write

$$\delta[n] = \begin{cases} 1 & n = 0 \\ 0 & \text{otherwise} \end{cases} \tag{2.31}$$

The graph of (2.30) is depicted in Fig. 2.90.

The relationship between the unit step function and the unit impulse function can be written as

$$\delta[n] = u[n] - u[n - 1] \tag{2.32}$$

or as

$$u[n] = \sum_{k = -\infty}^{n} \delta[k] \tag{2.33}$$

where first subtracting n from k and then replacing k by $-k$ we get

$$u[n] = \sum_{k = -\infty}^{n} \delta[k] \rightarrow u[n] = \sum_{k = -\infty}^{0} \delta[k + n] \rightarrow u[n] = \sum_{k = 0}^{\infty} \delta[-k + n]$$

Thus, the relationship between the unit step function and the unit impulse function can alternatively be written as

$$u[n] = \sum_{k = 0}^{\infty} \delta[n - k] \tag{2.34}$$

Fig. 2.90 Digital unit impulse function

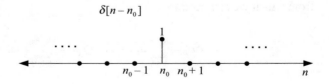

Fig. 2.91 Digital ramp function

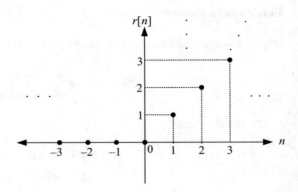

Ramp Function

Digital ramp function is defined as

$$r[n] = nu[n] \tag{2.35}$$

which can also be written as

$$r[n] = \begin{cases} n & n \geq 0 \\ 0 & \text{otherwise} \end{cases} \tag{2.36}$$

The graph of the ramp function is depicted in Fig. 2.91.

The relationship between the unit step function and the ramp function can be expressed using

$$u[n] = r[n] - r[n-1] \tag{2.37}$$

Another expression showing the relationship between the unit step function and the ramp function is

$$r[n] = \sum_{k=-\infty}^{n} u[k] \tag{2.38}$$

from which we can obtain

$$r[n] = \sum_{k=0}^{\infty} u[n-k] \rightarrow r[n] = \sum_{k=0}^{\infty} k\delta[n-k] \tag{2.39}$$

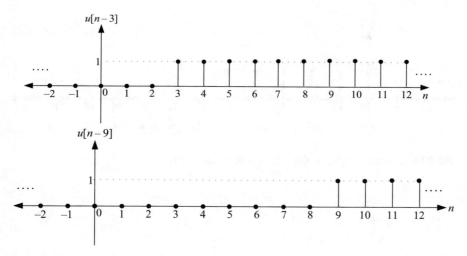

Fig. 2.92 Graphs of $u[n-3]$ and $u[n-9]$

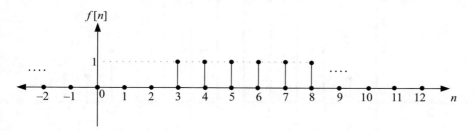

Fig. 2.93 Graph of $u[n-3] - u[n-9]$

Example 2.24 Draw the graph of $f[n] = u[n-3] - u[n-9]$.

Solution 2.24 The graphs of $u[n-3]$, $u[n-9]$ and $f[n]$ are shown in Figs. 2.92 and 2.93, respectively.

As it is seen from Fig. 2.93 when the subtraction $u[n-3] - u[n-9]$ is performed, overlapping parts cancel each other, and zero is produced. The function $f[n]$ is a rectangle between points 3 and 8.

Example 2.25 Draw the graph of $f[n] = u[n] - 2u[n-6] + u[n-11]$.

Solution 2.25 When the terms in $f[n]$ are inspected, it is seen that for $n \geq 6$ the function $-2u[n-6]$ is added to the unit step function $u[n]$, and this results in the amplitude value being equal to -1. For $n \geq 11$, the function $u[n-11]$ is further added to the previous summation, and this results in 0 amplitude value. The graph of $f[n]$ is shown in Fig. 2.94.

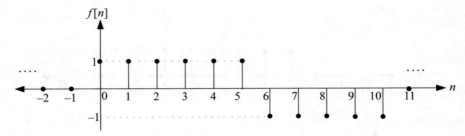

Fig. 2.94 The graph of $f[n] = u[n] - 2u[n - 6] + u[n - 11]$

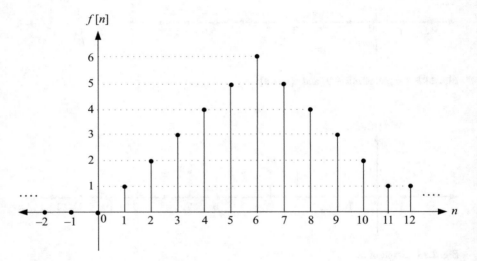

Fig. 2.95 The graph of $f[n] = r[n] - 2r[n - 6] + r[n - 11]$

Example 2.26 Draw the graph of $f[n] = r[n] - 2r[n - 6] + r[n - 11]$.

Solution 2.26 After point $n = 6$, the slope becomes equal to -1, and for $n \geq 11$, the function $f[n]$ becomes constant as depicted in Fig. 2.95.

Digital Exponential Function

Another function used in communication and signal processing is the exponential function which is defined as

$$f[n] = \begin{cases} ke^{-n} & n \geq 0 \\ 0 & \text{otherwise} \end{cases} \rightarrow f[n] = ke^{-n}u[n], \quad k \in R \qquad (2.40)$$

Example 2.27 Write the function shown in Fig. 2.96 in terms of the unit step functions.

Solution 2.27 First, let us identify the points where the function changes, and then considering these points, let us write the function as the sum of unit step functions. If we inspect the given function, we see that at points $n = 1$, $n = 3$ and $n = 6$ the amplitude of the function changes. Considering the function amplitude changes for $n = 1$ and $n = 3$, we can write the function for $1 \leq n \leq 5$ as $u[n - 1] - 2u[n - 3]$ as explained in Fig. 2.97.

Since the function has 0 value after $n = 6$, we can add the shifted function $u[n - 6]$ to the previous summation to achieve this goal as illustrated in Fig. 2.98. Thus, $f[n]$ can be written as $f[n] = u[n - 1] - 2u[n - 3] + u[n - 6]$.

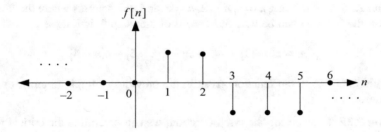

Fig. 2.96 Graph of $f[n]$ for Example 2.27

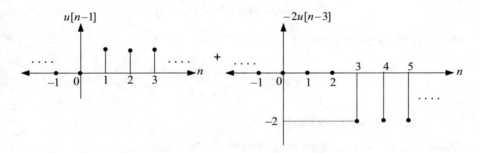

Fig. 2.97 Graphs of $u[n - 1]$ and $-2u[n - 3]$

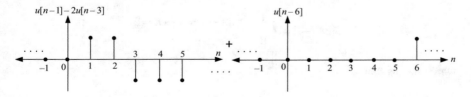

Fig. 2.98 Graph of $u[n - 1] - 2u[n - 3]$ and $u[n - 6]$.

Fig. 2.99 Graph of $f[n]$ for
Example 2.28

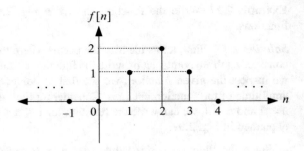

Example 2.28 Write the function shown in Fig. 2.99 in terms of the unit step functions.

Solution 2.28 Considering $n = 1, n = 2, n = 3$, and $n = 4$ points where the function changes, the function can be written in terms of unit step functions as

$$f[n] = u[n-1] + u[n-2] - u[n-3] - u[n-4]$$

Example 2.29 Write the function given in the previous example in terms of ramp functions.

Solution 2.29 Considering the ramp function, we can determine the critical points of the function as $n = 0, n = 2, n = 4$, and considering the slopes in the graph, we can write the function as

$$f[n] = r[n] - 2r[n-2] + r[n-4]$$

Exercise Write the function

$$f[n] = \begin{cases} -n & n \geq 0 \\ n+2 & n < 0 \end{cases}$$

in terms of discrete-time unit step functions.

Example 2.30 Write an open expression for $3\delta[2n - 4]$.

Solution 2.30 Using the definition of unit impulse function for $\delta[2n - 4]$, we can write

$$\delta[2n - 4] = \begin{cases} 1 & \text{if } 2n - 4 = 0 \rightarrow n = 2 \\ 0 & \text{otherwise} \end{cases}$$

which means that

$$3\delta[2n - 4] = \begin{cases} 3 & \text{if } n = 2 \\ 0 & \text{otherwise} \end{cases}$$

2.6   Electronic Circuits Used in Signal Generation

In this section, we will explain how some basic signal functions are produced with electronic circuits, and we will give sample circuit diagrams to explain the subject. Passive circuit elements are used to generate analog signals. Passive circuit elements consist of resistor, capacitor, and inductor, respectively. Basic signal functions can be generated using passive circuit elements as well as voltage or current sources. If we recall the basic signal functions, they were unit step signal, unit impulse signal, ramp signal, exponential signal, and sinusoidal signal.

2.6.1   Passive Circuit Elements

Inductor

An inductor is a passive circuit element used to store current, i.e., to store magnetic energy. The current value across the inductor does not show sudden changes; instead, it gradually decreases or increases (Fig. 2.100).

The relation between the current and the voltage on the inductor can be expressed using

$$v(t) = L\frac{di(t)}{dt} \rightarrow i(t) = \frac{1}{L}\int_{-\infty}^{t} v(\tau)d\tau \qquad (2.41)$$

Capacitor

A capacitor is one of the passive circuit elements used to store energy like an inductor. Voltage is stored on the capacitor, and voltage variations across the capacitor are not abrupt; voltage on a capacitor increases or decreases exponentially (Fig. 2.101).

Fig. 2.100 Symbolic representation of an inductor

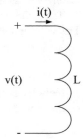

Fig. 2.101 Symbolic
representation of a capacitor

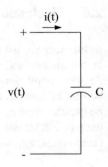

Fig. 2.102 Symbolic
representation of a resistor

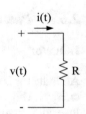

The current and voltage relationships on a capacitor can be described using

$$i(t) = C\frac{dv(t)}{dt} \rightarrow v(t) = \frac{1}{C} \int_{-\infty}^{t} i(\tau)d\tau \tag{2.42}$$

Resistor

The resistor is one of the passive circuit elements used to dissipate energy into the environment as heat and light. A resistor is not used for energy storage. As we mentioned earlier, inductors and capacitors are used to store energy. On the other hand, since these circuit elements have internal resistance, their stored energy is discharged as heat and light through the internal resistance (Fig. 2.102).

The value of the resistance can change depending on time, or it can be constant. The voltage-current relationship of the time-invariant resistor R is as

$$v(t) = Ri(t) \tag{2.43}$$

Now let us see how basic signals, i.e., unit step signal, impulse signal, sinusoidal signal, and exponential signal, are generated using passive circuit elements.

Fig. 2.103 Generation of the unit step function

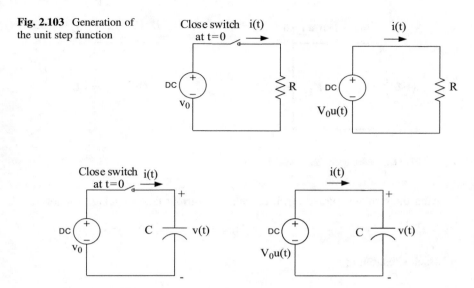

Fig. 2.104 Generation of the impulse function

Unit Step Function

The unit step function $u(t)$ takes values for $t \geq 0$; the operation of the unit step function can be considered as closing the switch in a circuit, and the unit step function can be used instead of a switch as illustrated in Fig. 2.103.

Impulse Function

To generate the impulse function $\delta(t)$, we can use the circuit in Fig. 2.104.

After the switch is closed, the voltage on the capacitor becomes $v_0 u(t)$, from which we calculate the current on the capacitor as

$$i(t) = C\frac{dv(t)}{dt} \rightarrow i(t) = C\frac{dv_0 u(t)}{dt} \rightarrow i(t) = Cv_0\frac{du(t)}{dt} \rightarrow i(t) = Cv_0\delta(t) \quad (2.44)$$

The current expression $i(t) = Cv_0\delta(t)$ includes the impulse function.

Sinusoidal Function

To generate a sinusoidal function, a capacitor with an initial voltage value of V_0 and an inductor is connected in parallel as shown in Fig. 2.105. After the switch is closed, assuming that the inductor and capacitor have no internal resistances, the voltage across the capacitor and the inductor will have sinusoidal characteristic.

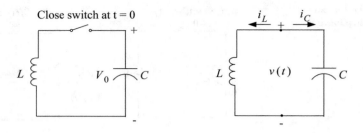

Fig. 2.105 Generation of the sinusoidal function

After the switch is turned off, if Kirchhoff's current law is applied, we get

$$i_L + i_C = 0$$

in which substituting

$$i_L = \frac{1}{L} \int_{-\infty}^{t} v(\tau)d\tau$$

and

$$i_C = C\frac{dv(t)}{dt}$$

we obtain

$$\frac{1}{L} \int_{-\infty}^{t} v(\tau)d\tau + C\frac{dv(t)}{dt} = 0 \quad t \geq 0. \tag{2.45}$$

By taking the derivative of (2.45) w.r.t. the time parameter "t," we obtain

$$\frac{1}{L}v(t) + C\frac{d^2v(t)}{dt^2} = 0, \quad t \geq 0 \tag{2.46}$$

whose solution can be found as

$$v(t) = V_0 \cos\left(\frac{t}{\sqrt{LC}}\right), \quad t \geq 0 \tag{2.47}$$

which is a sinusoidal function. The constant term V_0 in (2.47) is the voltage of the capacitor before the switch is closed.

Fig. 2.106 Generation of
the exponential function

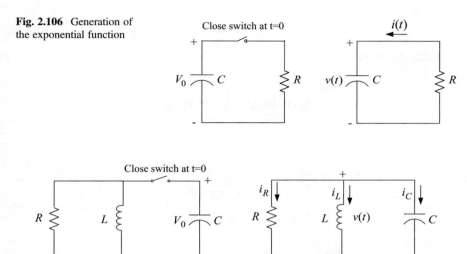

Fig. 2.107   Generation of the exponentially damped sinusoidal function

2.6.2   Exponential Function

The circuit in Fig. 2.106 can be used to generate the exponential function. In this circuit, a resistor is connected in series to the capacitor with an instantaneous voltage value V_0, and the switch is closed at $t = 0$. In this case, the voltage across the capacitor is discharged exponentially over the resistor. The voltage expression on the capacitor can be written as

$$v(t) = V_0 e^{-\frac{t}{RC}} \tag{2.48}$$

where RC is called the time constant of the circuit.

2.6.3   Exponentially Damped Sinusoidal Function

The circuit to generate the exponentially damped sine function is depicted in Fig. 2.107.

As it is seen from Fig. 2.107, an exponentially damped sine signal can be obtained from the circuit which is constructed by connecting the passive circuit elements, resistor, inductor, and capacitor in parallel. The instantaneous voltage value of the capacitor is V_0. If Kirchhoff's current law is applied after the switch is closed, we get

$$i_R + i_L + i_C = 0$$

in which substituting

$$i_R = \frac{v(t)}{R}, i_L = \frac{1}{L} \int\limits_{-\infty}^{t} v(\tau)d\tau$$

and

$$i_C = C\frac{dv(t)}{dt}$$

we obtain

$$\frac{v(t)}{R} + \frac{1}{L} \int\limits_{-\infty}^{t} v(\tau)d\tau + C\frac{dv(t)}{dt} = 0$$

whose derivative is taken as

$$\frac{1}{R}\frac{dv(t)}{dt} + \frac{v(t)}{L} + C\frac{d^2v(t)}{d^2t} = 0$$

whose solution can be found as

$$v(t) = V_0 e^{-\frac{t}{2RC}} \cos{(w_0 t)}, \quad t \geq 0 \tag{2.49}$$

where we have

$$w_0 = \sqrt{\frac{1}{LC} - \frac{1}{4C^2R^2}}$$

2.6.4   Linear and Time-Invariant Systems

The circuit shown in Fig. 2.108 can be given as an example of a linear time-invariant system. In this circuit, the system input is the voltage source $v(t)$, and the system output is the current $i(t)$.

Fig. 2.108 Linear and time-invariant system

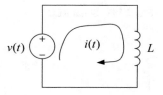

The relationship between input and output of the system can be expressed as

$$i(t) = \frac{1}{L} \int_{-\infty}^{t} v(\tau)d\tau \qquad (2.50)$$

Since the system is a linear system, for the summation of two different input voltages $v_1(t)$, $v_2(t)$, we can write

$$i(t) = \frac{1}{L} \int_{-\infty}^{t} (v_1(\tau) + v_2(\tau))d\tau$$

$$= \frac{1}{L} \int_{-\infty}^{t} v_1(\tau)d\tau + \frac{1}{L} \int_{-\infty}^{t} v_2(\tau)d\tau$$

$$= i_1(t) + i_2(t)$$

which verifies the linearity of the system. Now let us calculate the output of the system for the input $v(t - t_0)$ and examine whether the output is equal to $i(t - t_0)$ or not.

The system output for the input signal $v(t - t_0)$ can be calculated as

$$i_1(t) = \frac{1}{L} \int_{-\infty}^{t} v(\tau - t_0)d\tau$$

where if we make use of the parameter change, $\tau' = \tau - t_0$, we get

$$i_1(t) = \frac{1}{L} \int_{-\infty}^{t - t_0} v(\tau')d\tau' \qquad (2.51)$$

If (2.51) is compared to (2.50), we see that

$$i_1(t) = i(t - t_0)$$

which verifies that the system is a time-invariant system.

Fig. 2.109 Serial RLC circuit

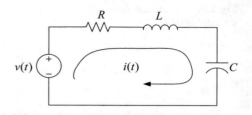

2.6.5 Representation of Linear and Time-Invariant System with Differential Equations

The relationship between the input and output of a linear and time-invariant system can be represented by differential or difference equations. The circuit in Fig. 2.109 can be given as an example of a linear and time-invariant system. In this circuit, $v(t)$ is the system input and $i(t)$ is the system output.

If Kirchhoff's voltage law is applied to the circuit in Fig. 2.109, we get

$$v(t) = Ri(t) + L\frac{di(t)}{dt} + \frac{1}{C}\int_{-\infty}^{t} i(\tau)d\tau$$

where taking the derivative of the left and the right sides, we obtain

$$\frac{dv(t)}{dt} = R\frac{di(t)}{dt} + L\frac{d^2i(t)}{dt^2} + \frac{1}{C}i(t)$$

which is a differential equation whose solution can be obtained as

$$i(t) = Ae^{-\alpha t}\cos{(wt + \theta)}, \quad \alpha = \frac{R}{2L} \tag{2.52}$$

Problems

1. The graph of the function $f(t)$ is depicted in Fig. P1.1.
   Draw the graphs of the following functions:

   (a) $-f(t)$
   (b) $f(2t)$
   (c) $-f(2t)$
   (d) $f(t-1)$
   (e) $f(2t-1)$
   (f) $-2f(2t-1)$
   (g) $f(t-2)$

(h) $-f(-t-2)$
(i) $f(3t+6)$
(j) $f(-3t+6)$

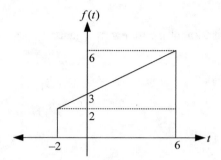

Fig. P1.1 The graph of $f(t)$ for Problem 1

2. The graph of the function $f[n]$ is depicted in Fig. P1.2.
   Draw the graphs of the following functions:

(a) $f[-n]$
(b) $f[-2n]$
(c) $f[2n]$
(d) $f[n-1]$
(e) $f[2n-1]$
(f) $f[-2n-1]$
(g) $f[n+1]$
(h) $f\left[\frac{n}{3}+1\right]$
(i) $f\left[-\frac{n}{3}+1\right]$
(j) $f\left[-\frac{n}{3}-2\right]$
(k) $-2f[3n]$
(l) $-3f\left[-\frac{n}{2}\right]$

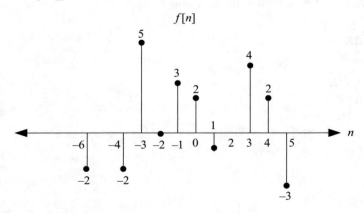

Fig. P1.2 The graph of $f[n]$ for Problem 2

3. $f[n]$ in vector form is given as

$$f[n] = \left[\begin{array}{ccccccccccc} -2 & -4 & 2.5 & 3 & \underset{n=0}{4} & 6 & 5 & -3 & -2 & 1 & 0 & 2 \end{array} \right]$$

Find the vector form of the following signals:

(a) $f\left[\frac{n}{2}\right]$
(b) $f[n-1]$
(c) $f[-n-1]$
(d) $f[2n-1]$
(e) $f[-2n-1]$
(f) $2f[-2n-1]$
(g) $-0.5f[-2n-4]$

4. Draw the graphs of the following signals:

(a) $\delta(2t)$
(b) $\delta(2t-1)$
(c) $\delta(2t+1) - \delta(2t-3)$
(d) $u(t-2) + u(t-4) - 2u(t-6)$
(e) $r(t+1) - 2r(t-1) + r(t-2)$
(f) $\delta[2n]$
(g) $\delta[2n+4] + \delta[2n-2]$
(h) $\delta\left[\frac{n}{3}-2\right] + \delta\left[\frac{n}{2}+1\right]$

5. $f[n]$ is given as

$$f[n] = \left[\begin{array}{cccccccccccc} \underset{n=0}{1} & -1 & 0 & 2 & -1 & 3 & -1 & 0 & 0 & -1 & 0 & 2 \end{array} \right]$$

Find the vector form of the following signals:

(a) $f[n-1]$
(b) $f[2n-1]$
(c) $f[2n]$
(d) $\sqrt{f\left[\frac{n}{2}\right]}$
(e) $f[-2n]$
(f) $-3f[-2n+2]$
(g) $\frac{1}{2}f[-2n+3]$
(h) $f^2\left[-\frac{n}{2}-3\right]$

6. The digital signal $f[n]$ is defined as

$$f[n] = \delta[n+4] + 2\delta[n+2] + \delta[n] - 3\delta[n-2] + 5\delta[n-6] - \delta[n-9]$$

Draw the graph of $f[n]$, and draw the graphs of the following signals:

(a) $f[2n]$
(b) $f[2n+2]$
(c) $f\left[\frac{n}{2}\right]$
(d) $-3f[3n+3]$
(e) $2f\left[\frac{n}{2}+4\right]$
(f) $\sqrt{\frac{1}{5}f[9n-3]}$

7. The continuous-time signal $f(t) = \sin(100\pi t) + \sin(1000\pi t)$ is sampled. What should be the minimum value of the sampling frequency such that the analog signal can be perfectly reconstructed from its samples? Using the sampling frequency, write the mathematical form of the digital signal $f[n]$ obtained from continuous-time signal via sampling operation.

8. Write the $f(t)$ given in problem (1) in terms of unit step and ramp functions. Next, calculate $\frac{df(t)}{dt}$ and draw its graph. Write mathematical expressions for $f\left(\frac{t^2}{3}\right)$ and its derivative function.

9. The graph of $f(t)$ is depicted in Fig. P1.3. Draw the graph of

$$g(t) = \sum_k f\left(\frac{t-k2\pi}{T}\right)$$

for $T = 1$, $T = 2$, and $T = 4$.

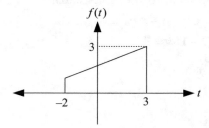

Fig. P1.3 The graph of $f(t)$ for Problem 9

10. Draw the graphs of the following functions: $r[n]$ is the ramp function:

(a) $f[n] = r[n] + r[n-4] - r[n-8]$
(b) $g[n] = r[n] + r[n-4] - 2r[n-8]$

11. Write the function, whose graph is depicted in Fig. P1.4, in terms of ramp functions.

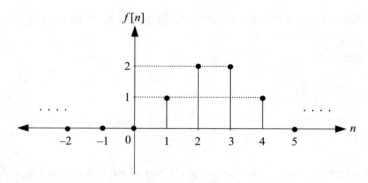

Fig. P1.4 The graphs of functions for Problem 11

12. Draw the graphs of the following functions:

   (a) $u[2n]$
   (b) $u[3n]$
   (c) $r[2n]$
   (d) $r[3n]$

13. Find the periods of the following functions:

   (a) $\sin(t)$
   (b) $\cos\left(\frac{\pi}{3}t.\right)$

14. Calculate the results of the following operations:

   (a) $\int_{t=-\infty}^{\infty}(t^2+1)\delta(t-1)dt=?$
   (b) $\int_{t=2}^{\infty}(t^2+1)\delta(t-1)dt=?$
   (c) $\delta(t-1)\delta(t+3)=?$
   (d) $\delta(t-1)u(t)=?$

Chapter 3
Energy, Power, Convolution, and Systems

In this chapter, we will first deal with the calculation of energy and power of signals. The signals that we will discuss in this chapter are deterministic signals, and they do not have random characteristics. The calculation of power expressions of random signals is beyond the scope of this book. Knowing the energy and power of a signal is an important issue in communication. In particular, many performance criteria are based on the signal power-to-noise power ratio at the receiver. Energy is the total power dissipated by a signal in a given time interval. The power of a signal can be considered as the amount of energy dissipated per unit of time. We can perceive and measure energy as heat and light as it transforms from one form to another. For some signals, the power calculations can be performed, but their energies cannot be calculated, whereas for some other signals, the energies can be calculated, but powers cannot be calculated. In the following sections, we will give examples of these signals and group these signals under different names. Energy and power calculations can be done both in the time domain and in the frequency domain. For some signals, energy calculations can be difficult in the time domain, whereas energy or power calculations in the frequency domain can be made more easily. In this section, we will calculate energy and power only in the time domain. We will consider energy and power calculations in the frequency domain after we study the Fourier transform topic.

The second topic we will study in this chapter is the convolution of signals. Systems can be regarded as electronic circuits that process signals and generate new signals from them. Some systems are described mathematically. We cannot say that every mathematically defined system has a physical counterpart. Some mathematical definitions remain only theoretical; they cannot be applied in practice. In order for the mathematically defined systems to be applied in practice, these systems must have some properties. One of these properties is causality. Another important property is stability. Unstable systems are not preferred in practical applications. During the theoretical design of any system, attention should be paid to its stability. Another property of a system is memory. While for some systems having memory can be a critical requirement, for others not having memory can be an important

O. Gazi, *Principles of Signals and Systems*, https://doi.org/10.1007/978-3-031-17789-7_3

criterion. Linearity and time invariance are the remaining properties that systems can
have. Since it is easier to process linear and time-invariant systems, it is preferred
that the systems we design are linear and time-invariant. If the impulse responses of
linear and time-invariant systems are known, the output of the system for any
random input can be calculated by the convolution operation.

3.1 Energy and Power in Signals

3.1.1 Energy and Power in Continuous-Time Signals

Instantaneous Power

The instantaneous power of $x(t)$ is defined as

$$p(t) = x^2(t) \tag{3.1}$$

Total Energy

Using instantaneous power, total energy can be calculated as

$$E_x = \lim_{\tau \to \infty} \int_{-\frac{\tau}{2}}^{\frac{\tau}{2}} p(t)dt \tag{3.2}$$

Average Power

Average power of a signal whose total energy is known is calculated using

$$P = \lim_{\tau \to \infty} \frac{1}{\tau} \int_{-\frac{\tau}{2}}^{\frac{\tau}{2}} p(t)dt \tag{3.3}$$

For periodic signals, the average power is calculated using

$$P = \frac{1}{T} \int_{-\frac{T}{2}}^{\frac{T}{2}} x^2(t)dt$$

where T is the period of $x(t)$ such that $x(t) = x(t + T)$. For periodic signals, root-
mean-square is defined as

$$P_{rms} = \sqrt{\frac{1}{T} \int_{-\frac{T}{2}}^{\frac{T}{2}} x^2(t)dt} \tag{3.4}$$

Example 3.1 Find the instantaneous, average power, and energy of

$$x(t) = u(t).$$

Solution 3.1 The instantaneous power can be calculated using $p(t) = x^2(t)$ as

$$\begin{aligned} p(t) &= u^2(t) \\ &= u(t)u(t) \\ &= u(t) \end{aligned}$$

which can be used for the calculation of the average power as in

$$\begin{aligned} P &= \lim_{\tau \to \infty} \frac{1}{\tau} \int_{-\frac{\tau}{2}}^{\frac{\tau}{2}} p(t)dt \\ &= \lim_{\tau \to \infty} \frac{1}{\tau} \int_{-\frac{\tau}{2}}^{\frac{\tau}{2}} u(t)dt \\ &= \lim_{\tau \to \infty} \frac{1}{\tau} \int_{0}^{\frac{\tau}{2}} u(t)dt \\ &= \lim_{\tau \to \infty} \frac{\frac{\tau}{2}}{\tau} \\ &= \frac{1}{2} \end{aligned}$$

While performing the integration, note that the lower bound of the integral is 0, since the unit step function takes values only for positive t values. The total energy of the signal can be calculated as

$$E_x = \lim_{\tau \to \infty} \int_{-\frac{\tau}{2}}^{\frac{\tau}{2}} p(t)dt$$

$$= \lim_{\tau \to \infty} \int_{0}^{\frac{\tau}{2}} u(t)dt$$

$$= \lim_{\tau \to \infty} \frac{\tau}{2}$$

$$= \infty$$

Hence, we see that the instantaneous and average power of the unit step function can be calculated, but the energy of the signal is infinite, which can be understood from the graph of the unit step function. Since the unit step function is a rectangle with amplitude "1" going to infinity on the positive axis, its square is the same, and the area under its square is infinite. On the other hand, since the amplitude remains the same, the instantaneous power and the average power are constant numbers.

Signals whose energies cannot be calculated, but whose powers can be calculated, are called *power* signals. On the other hand, signals whose powers cannot be calculated, but whose energies can be calculated, are called *energy* signals.

Example 3.2 Calculate the instantaneous power, average power, and energy of the $x(t)$ signal is given as

$$x(t) = \begin{cases} t & 0 \leq t \leq 2 \\ 3 - t & 2 \leq t \leq 4 \\ 0 & \text{otherwise} \end{cases} \tag{3.5}$$

Solution 3.2 Instantaneous power can be calculated using $p(t) = x^2(t)$ as

$$p(t) = x^2(t) = \begin{cases} t^2 & 0 \leq t \leq 2 \\ (3 - t)^2 & 2 \leq t \leq 4 \\ 0 & \text{otherwise} \end{cases}$$

The average signal power of the signal can be calculated as

$$P = \lim_{\tau \to \infty} \frac{1}{\tau} \int_{-\frac{\tau}{2}}^{\frac{\tau}{2}} p(t)dt$$

$$= \lim_{\tau \to \infty} \frac{1}{\tau} \left(\int_{0}^{2} t^2 dt + \int_{2}^{4} (3-t)^2 dt \right)$$

$$= 0$$

We can calculate the energy of the signal as in

$$E_x = \lim_{\tau \to \infty} \int_{-\frac{\tau}{2}}^{\frac{\tau}{2}} p(t)dt$$

$$= \int_{0}^{2} t^2 dt + \int_{2}^{4} (3-t)^2 dt$$

$$= \left(\frac{t^3}{3} \Big|_0^2 + \left(- \frac{(3-t)^3}{3} \Big|_2^4 \right) = \frac{8}{3} + \frac{2}{3} = \frac{10}{3} \right.$$

The signal does not have average power; however, it has finite energy. We call such signals energy signals.

Note In this book, when the signal power is mentioned, we will assume that the average power of the signal is considered, not instantaneous power unless otherwise indicated.

Example 3.3 Calculate the energy and power of the signal

$$x(t) = A \sin \left(\frac{2\pi}{T} t \right) \tag{3.6}$$

The period of the sine signal is T.

Solution 3.3 The energy of the sine signal is found by calculating the area under the function

$$x^2(t) = A^2 \sin^2 \left(\frac{2\pi}{T} t \right)$$

Since the mathematical expression $x^2(t) = A^2 \sin^2\left(\frac{2\pi}{T}t\right)$ always takes positive values, its integral is infinite. That is, the energy of $x(t)$ is infinite. The power of the periodic signal is calculated using its one period as in

$$
\begin{aligned}
P &= \frac{1}{T} \int_{-\frac{T}{2}}^{\frac{T}{2}} x^2(t)dt \\[2mm]
&= \frac{1}{T} \int_{-\frac{T}{2}}^{\frac{T}{2}} A^2 \sin^2\left(\frac{2\pi}{T}t\right)dt \\[2mm]
&= \frac{A^2}{T} \int_{-\frac{T}{2}}^{\frac{T}{2}} \frac{\left(1 - \cos\left(\frac{4\pi}{T}t\right)\right)}{2} dt \\[2mm]
&= \frac{A^2}{2T} \int_{-\frac{T}{2}}^{\frac{T}{2}} \left(1 - \cos\left(\frac{4\pi}{T}t\right)\right)dt \\[2mm]
&= \frac{A^2}{2} - 0 \\[2mm]
&= \frac{A^2}{2}
\end{aligned}
$$

from which it is seen that the average power of the sine signal is equal to half of the square of its maximum amplitude.

3.1.2 Energies and Powers of Discrete-Time Signals

In this section, we make the energy and power definitions for digital, $x[n]$, signals.

Instantaneous Power

The instantaneous power of the digital signal $x[n]$ is defined as

$$p[n] = x^2[n] \tag{3.7}$$

Total Energy

The total energy of the digital signal $x[n]$ is calculated as

$$E_x = \sum_{n=-\infty}^{\infty} x^2[n] \tag{3.8}$$

Average Power

The average power of the digital signal $x[n]$ is calculated using

$$P = \lim_{M \to \infty} \frac{1}{2M+1} \sum_{n=-M}^{M} x^2[n] \tag{3.9}$$

If $x[n]$ is a periodic signal with period N, then its average power is calculated over its one period as

$$P = \frac{1}{N} \sum_{n=0}^{N-1} x^2[n] \tag{3.10}$$

Example 3.4 Find the instantaneous power, average power, and energy of

$$x[n] = (-1)^n$$

Solution 3.4 The instantaneous power is calculated as

$$\begin{aligned} p[n] &= x^2[n] \\ &= (-1)^{2n} \\ &= 1 \end{aligned}$$

The average power of the signal is calculated as in

$$\begin{aligned} P &= \lim_{M \to \infty} \frac{1}{2M+1} \sum_{n=-M}^{M} x^2[n] \\ &= \lim_{M \to \infty} \frac{1}{2M+1} \sum_{n=-M}^{M} 1 \\ &= \lim_{M \to \infty} \frac{2M+1}{2M+1} \\ &= 1 \end{aligned}$$

The signal energy obtained as

$$E_x = \sum_{n=-\infty}^{\infty} x^2[n]$$
$$= \sum_{n=-\infty}^{\infty} 1$$
$$= \infty$$

From the obtained results, we see that the signal has finite power but infinite energy. Thus, this signal is a power signal.

Exercise Calculate the instantaneous power, the average power, and the energy of $x[n] = \cos(3\pi n)$.

3.2 Convolution

Convolution should be learned very well because it is a subject that we will encounter frequently in signal processing and communication. We will first deal with finding the convolutions of continuous signals and then the convolutions of discrete signals. Before starting to study the topic of convolution, we advise the reader to read the definitions and properties of continuous and discrete signals from the relevant chapters of the book.

3.2.1 Convolution of Continuous-Time Signals

Let $f(t)$ and $g(t)$ be two continuous-time functions.

The convolution of these two functions is indicated as $f(t) * g(t)$, and the convolution is defined as

$$h(t) = f(t) * g(t)$$
$$= \int_{-\infty}^{\infty} f(\tau)g(t-\tau)d\tau \tag{3.11}$$

where the parameter τ disappears after the evaluation of the integral. Any letter can also be used for the place of τ. To calculate the convolution of two functions, sometimes mathematical methods are easier to deal with, and at other times, graphical calculations can be easier as well. We can decide which method we should use by inspecting the question and using our own experience.

Example 3.5 Calculate $h(t) = f(t) * f(t)$ where $f(t) = u(t)$.

Solution 3.5-(1) The convolution operation can be calculated as

$$
\begin{aligned}
h(t) &= f(t) * f(t) \\
&= \int_{-\infty}^{\infty} f(\tau) f(t - \tau) d\tau \\
&= \int_{-\infty}^{\infty} u(\tau) u(t - \tau) d\tau
\end{aligned}
$$

where writing $u(\tau)$ and $u(t - \tau)$ as

$$
u(\tau) = \begin{cases} 1 & \tau \geq 0 \\ 0 & \text{otherwise} \end{cases}
\qquad
u(t - \tau) = \begin{cases} 1 & t - \tau \geq 0, \; \rightarrow t \geq \tau \\ 0 & \text{otherwise} \end{cases}
$$

we get

$$
u(\tau) u(t - \tau) = \begin{cases} 1 & t \geq \tau \geq 0 \\ 0 & \text{otherwise} \end{cases}
$$

which is the integrand expression and the integral is evaluated as

$$
\begin{aligned}
h(t) &= \int_{0}^{t} 1 \times 1 d\tau \quad t \geq 0 \\
&= t \qquad\qquad\quad t \geq 0
\end{aligned}
$$

which is a ramp function, i.e., $h(t) = tu(t) \rightarrow r(t)$. Thus, the convolution of the two unit step functions gives us the ramp function. That is, we have

$$
u(t) * u(t) = r(t) \tag{3.12}
$$

Solution 3.5-(2) Now, let us solve the same question graphically without using a mathematical approach.

First, let us draw the graphs of the functions $u(\tau)$ and $u(t - \tau)$. Then, let us find their product, and calculate the integral of the product.

The graphs of $u(\tau)$, $u(t - \tau)$, and $u(\tau) u(t - \tau)$ are depicted in Fig. 3.1. It is seen from Fig. 3.1 that the total area under $u(\tau) u(t - \tau)$ is $t. 1 = t$ which is the integral of $u(\tau) u(t - \tau)$ over the interval $[0 \; t]$. Hence, we can write that

$$
u(t) * u(t) = t, \quad t > 0
$$

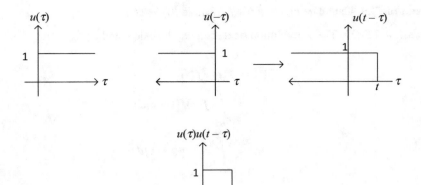

Fig. 3.1 Graphs of $u(\tau)$, $u(-\tau)$, $u(t-\tau)$ and $u(\tau)u(t-\tau)$

When we inspect the graph of $u(\tau)u(t-\tau)$, we see that the total area under $u(\tau)$ $u(t-\tau)$ depends on t. As the value of t increases, the area of the rectangle increases.

Property Convolution is commutative, that is, the convolution of $f(t)$ and $g(t)$ can be calculated using either $f(t) * g(t)$ or $g(t) * f(t)$, i.e., we have

$$f(t) * g(t) = g(t) * f(t)$$

Proof The convolution of two functions is calculated as

$$h(t) = f(t) * g(t)$$
$$= \int_{-\infty}^{\infty} f(\tau)g(t-\tau)d\tau$$

where making use of the parameter change as

$$t - \tau = s \rightarrow \tau = t - s, d\tau = -ds$$

we obtain

$$\int_{-\infty}^{\infty} f(\tau)g(t-\tau)d\tau = -\int_{+\infty}^{-\infty} f(t-s)g(s)ds$$
$$= \int_{-\infty}^{\infty} g(s)f(t-s)ds$$
$$= g(t) * f(t)$$

which indicates the commutative property of the convolution.

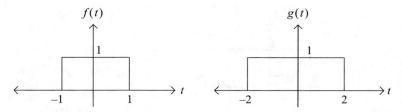

Fig. 3.2 Graphs of $f(t)$ and $g(t)$ for Example 3.6

Exercise Show that convolution has the distributive property, i.e., we have

$$f(t) * (g(t) + h(t)) = f(t) * g(t) + f(t) * h(t)$$

Example 3.6 Find the convolution of two functions whose graphs are depicted in Fig. 3.2.

Solution 3.6 We can express the functions shown in Fig. 3.2 in terms of unit step functions and calculate the convolution algebraically.

In order to find the convolution, we can write the functions in terms of unit step functions and calculate the convolution mathematically. The method we will follow will be the graphical approach method. We should decide whether we use $g(t) * f(t)$ or $f(t) * g(t)$. In fact, we should choose the one which is easier to calculate. For some problems, the graphical calculation of $f(t) * g(t)$ may be much simpler than $g(t) * f(t)$ and vice versa, even though there is not much difference between the two for this question.

Now let us continue with the formula

$$g(t) * f(t) = \int_{-\infty}^{\infty} g(\tau) f(t - \tau) d\tau \qquad (3.13)$$

Examine the graphs shown in Fig. 3.3 carefully, and try to understand how the overlapping parts of the $g(\tau)$ and $f(t - \tau)$ change according to the variable t.

Considering the different intervals used for t in Fig. 3.3 and integration results, we can write the convolution result as in

$$h(t) = g(t) * f(t)$$
$$= \begin{cases} 0 & t < -3 \\ t + 3 & -3 \leq t < -1 \\ 2 & -1 \leq t < 1 \\ 3 - t & 1 \leq t < 3 \\ 0 & 3 \leq t \end{cases}$$

whose graph is depicted in Fig. 3.4.

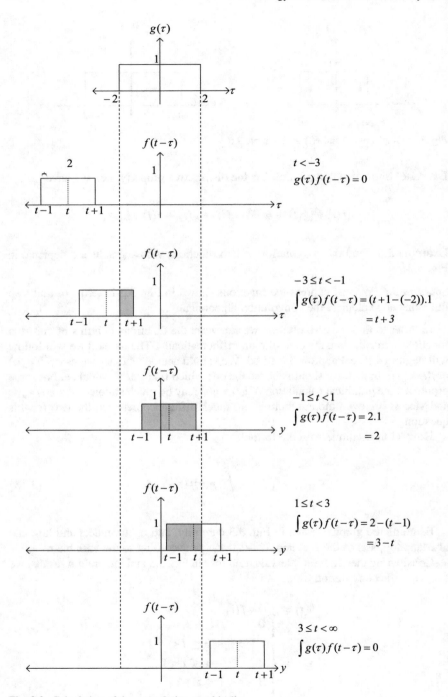

Fig. 3.3 Calculation of the convolution graphically

Fig. 3.4 Result of the convolution

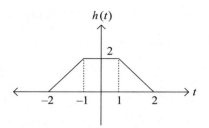

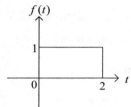

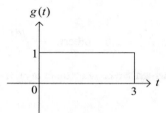

Fig. 3.5 Graphs of $f(t)$ and $g(t)$ for exercise

Fig. 3.6 Graphs of $f(t)$ and $g(t)$ for Example 3.7

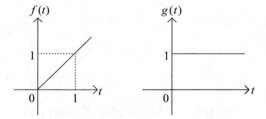

Exercise Find the convolution of the functions whose graphs are depicted in Fig. 3.5.

Example 3.7 Find the convolution of the functions shown in Fig. 3.6 using both graphical and mathematical approaches separately.

Solution 3.7 First, let us calculate the convolution integral mathematically and then graphically. The functions can be written as

$$f(t) = tu(t) \quad g(t) = u(t)$$

whose convolution can be calculated as

$$h(t) = f(t) * g(t)$$

$$= \int_{-\infty}^{\infty} f(\tau)g(t-\tau)d\tau$$

$$= \int_{-\infty}^{\infty} \tau u(\tau)u(t-\tau)d\tau$$

where $u(\tau)$ and $u(t-\tau)$ can be written as

$$u(\tau) = \begin{cases} 1 & \tau \geq 0 \\ 0 & \text{otherwise} \end{cases} \qquad u(t-\tau) = \begin{cases} 1 & t-\tau \geq 0, \ \rightarrow t \geq \tau \\ 0 & \text{otherwise} \end{cases}$$

From which we can calculate $u(\tau)u(t-\tau)$ as in

$$u(\tau)u(t-\tau) = \begin{cases} 1 & t \geq \tau \geq 0 \\ 0 & \text{otherwise} \end{cases}$$

and in the sequel, the convolution integral can be evaluated as

$$h(t) = \int_0^t \tau d\tau$$

$$= \frac{t^2}{2}, \quad t \geq 0$$

The graphs of the functions $f(\tau)$ and $g(t-\tau)$ and $f(\tau)g(t-\tau)$ can be drawn as in Fig. 3.7.

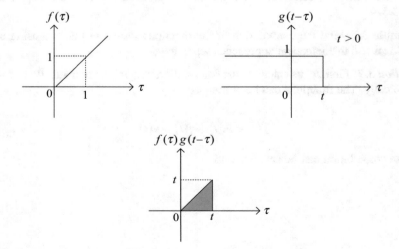

Fig. 3.7 The graphs of the functions $f(\tau)$ and $g(t-\tau)$ and $f(\tau)g(t-\tau)$

It is seen in Fig. 3.7 that the total area under the function $f(\tau)g(t - \tau)$ equals $\frac{t^2}{2}$ which is the value of the convolution integral.

Example 3.8 $u(t - a) * u(t - b) = ?$

Solution 3.8 Using the definition of convolution integration, we can calculate the convolution as

$$f(t) = u(t - a), \quad g(t) = u(t - b)$$
$$h(t) = f(t) * g(t)$$
$$= \int_{\tau = -\infty}^{\infty} f(\tau)g(t - \tau)d\tau$$
$$= \int_{\tau = -\infty}^{\infty} u(\tau - a)u(t - \tau - b)d\tau$$

$$u(\tau - a) = \begin{cases} 1 & \tau \geq a \\ 0 & \text{otherwise} \end{cases} \qquad u(t - \tau - b) = \begin{cases} 1 & \tau \leq t - b \\ 0 & \text{otherwise} \end{cases}$$

$$u(\tau - a)u(t - \tau - b) = \begin{cases} 1 & a \leq \tau \leq t - b \\ 0 & \text{otherwise} \end{cases}$$

$$h(t) = \int_{a}^{t-b} 1 d\tau$$
$$= \begin{cases} t - a - b & t \geq a + b \\ 0 & \text{otherwise} \end{cases}$$

It is possible to write $h(t)$ in terms of shifted unit step function or shifted ramp function as in

$$h(t) = (t - a - b)u(t - a - b)$$
$$= r(t - a - b)$$

We can directly use the result we have obtained to solve some kinds of problems. In order to use this result, the function must be writeable in terms of unit step functions.

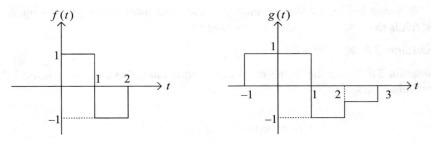

Fig. 3.8 Graphs of $f(t)$ and $g(t)$ for Example 3.9

Example 3.9 Find the convolution of the functions depicted in Fig. 3.8.

Solution 3.9 Let us write the functions $f(t)$ and $g(t)$ in terms of the unit step functions as

$$f(t) = u(t) - 2u(t-1) + u(t-2)$$
$$g(t) = u(t+1) - 2u(t-1) + u(t-2) + u(t-3)$$

Now we can calculate the convolution. Using the distributive property of the convolution, we can write

$$
\begin{aligned}
f(t) * g(t) &= (u(t) - 2u(t-1) + u(t-2)) \\
&\quad *(u(t+1) - 2u(t-1) + u(t-2) + u(t-3)) \\
&= u(t) * (u(t+1) - 2u(t-1) + u(t-2) + u(t-3)) \\
&\quad + -2u(t-1) * (u(t+1) - 2u(t-1) + u(t-2) + u(t-3)) \\
&\quad + u(t-2) * (u(t+1) - 2u(t-1) + u(t-2) + u(t-3))
\end{aligned}
$$

where using the distributive property again, we obtain

$$
\begin{aligned}
f(t) * g(t) &= u(t) * u(t+1) - 2u(t) * u(t-1) + u(t) * u(t-2) + u(t) \\
&\quad * u(t-3) - 2u(t-1) * u(t+1) + 4u(t-1) * u(t-1) \\
&\quad - 2u(t-1) * u(t-2) - 2u(t-1) * u(t-3) + u(t-2) \\
&\quad * u(t+1) - 2u(t-2) * u(t-1) + u(t-2) * u(t-2) \\
&\quad + u(t-2) * u(t-3)
\end{aligned}
$$

where employing the result we obtained in the previous problem, we get

$$
\begin{aligned}
f(t) * g(t) &= r(t+1) - 2r(t-1) + r(t-2) + r(t-3) - 2r(t) \\
&\quad + 4r(t-2) - 2r(t-3) - 2r(t-4) + r(t-1) - 2r(t-3) \\
&\quad + r(t-4) + r(t-5)
\end{aligned}
$$

which is further simplified as

$$f(t) * g(t) = -2r(t) + r(t+1) - r(t-1) + 5r(t-2) - 3r(t-3)$$
$$- r(t-4) + r(t-5)$$

Exercise $r(t-a) * r(t-b) = ?$ where a, b are two real numbers.

3.2.2 Impulse Function and Convolution

A frequently used function in convolution is the impulse function. The impulse function can be defined as

$$\delta(t) = \begin{cases} \infty & t = 0 \\ 0 & \text{otherwise} \end{cases} \qquad \delta(t - t_0) = \begin{cases} \infty & t = t_0 \\ 0 & \text{otherwise} \end{cases} \qquad (3.14)$$

which satisfy

$$\int_{-\infty}^{\infty} \delta(t)dt = 1 \qquad \int_{-\infty}^{\infty} \delta(t - t_0)dt = 1 \qquad (3.15)$$

Any function can be written as the convolution of the impulse function, namely:

$$f(t) = f(t) * \delta(t)$$
$$= \int_{-\infty}^{\infty} f(\tau)\delta(t - \tau)d\tau$$

The impulse function has the properties:

$$f(t) * \delta(t - t_0) = f(t - t_0) \qquad \int_{t=-\infty}^{\infty} f(t) * \delta(t - t_0)dt = f(t_0) \qquad (3.16)$$

Example 3.10 $\delta(t+1) * \delta(t-1) = ?$

Solution 3.10 To solve this problem, we will use the property

$$f(t) * \delta(t - t_0) = f(t - t_0)$$

Let $f(t) = \delta(t+1)$, and then we can write

$$\delta(t+1) * \delta(t-1) = \delta(t+1-1)$$
$$= \delta(t)$$

Property Assume that the function $g(t)$ has n roots, and these roots can be evaluated from the function at time instants $t_0, t_1, \ldots, t_{n-1}$. The function $\delta(g(t))$ can be written in terms of $\delta(t)$ using the roots of $g(t)$ as in

$$\delta(g(t)) = \sum_{i=0}^{n-1} \frac{\delta(t-t_i)}{|g'(t_i)|} \tag{3.17}$$

where $g'(t_i)$ is the derivative of $g(t)$ evaluated at t_i. Using (3.17), we get the expression

$$\delta(at-b) = \frac{\delta\left(t-\frac{b}{a}\right)}{|a|} \tag{3.18}$$

Again using (3.17), we get the equality

$$\int f(t)\delta(g(t))dt = \sum_{i=0}^{n-1} \frac{f(t_i)}{|g'(t_i)|} \tag{3.19}$$

Example 3.11 $\delta(2t-1) = ?$

Solution 3.11 Let $g(t) = 2t-1$, the root of $g(t)$ can be found from $g(t) = 0$ as $t_0 = \frac{1}{2}$, and then using (3.17), we obtain

$$\delta(2t-1) = \frac{\delta\left(t-\frac{1}{2}\right)}{2}$$

Example 3.12 $\delta(t^2-1) = ?$

Solution 3.12 First finding the roots of $t^2 - 1 = 0$ and then using (3.17), we write

$$\delta(t^2-1) = \frac{1}{2}[\delta(t-1) + \delta(t+1)]$$

Example 3.13 $\delta(t^2-1) * \delta(t^2-1) = ?$

Solution 3.13 First, let us expand $\delta(t^2-1)$ as in

$$\delta(t^2 - 1) = \frac{1}{2}[\delta(t-1) + \delta(t+1)]$$

and then using the distributive and linearity properties of the convolution, we can calculate the convolution in question as

$$\frac{1}{2}(\delta(t-1) + \delta(t+1)) * \frac{1}{2}(\delta(t-1) + \delta(t+1))$$
$$= \frac{1}{4}\delta(t-1) * \delta(t-1) + \frac{1}{4}\delta(t-1) * \delta(t+1)$$
$$+ \frac{1}{4}\delta(t+1) * \delta(t-1) + \frac{1}{4}\delta(t+1) * \delta(t+1)$$
$$= \frac{1}{4}(\delta(t-2) + 2\delta(t) + \delta(t+2))$$

Hence, we obtained

$$\delta(t^2 - 1) * \delta(t^2 - 1) = \frac{1}{4}(\delta(t-2) + 2\delta(t) + \delta(t+2))$$

3.2.3 Convolution of Digital Signals

Digital signals are discrete-time signals. Every digital signal is a discrete-time signal, but the reverse is not true. For instance, the signal

$$f(t) = \delta(t-1) + \delta(t+1)$$

is a discrete-time signal, but it is not a digital signal. On the other hand, the signal

$$f[n] = \delta[n-1] + \delta[n+1]$$

is a discrete signal in time, and it is also a digital signal.

Discrete signals are represented as a sequence of numbers. Each number in the sequence represents the amplitude of the signal at a point in time. The time axis values increase one by one. It is critical to know the moment of time zero at the time axis for reference.

In Fig. 3.9, the graph of a discrete-time signal, a digital signal, is given.

Fig. 3.9 A digital signal

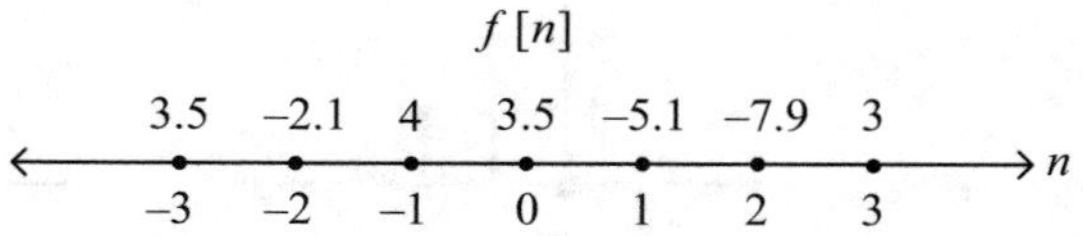

$$f[n]$$

3.5	−2.1	4	3.5	−5.1	−7.9	3
−3	−2	−1	0	1	2	3

The digital signal shown in Fig. 3.9 can be represented in vector form as

$$f[n] = \begin{bmatrix} 3.5 & -2.1 & 4.0 & \underset{n=0}{\underbrace{3.5}} & -5.1 & -7.9 & 3 \end{bmatrix}$$

where reference point $n = 0$ is indicated clearly. The time indices to the left of the reference point decrease one by one; on the other hand, the time indices to the right of the reference point increase one by one.

3.2.4 Convolution of Digital Signals

Let $f[n]$ and $g[n]$ be two digital signals. The convolution of these two digital signals is calculated as

$$\begin{aligned} h[n] &= f[n] * g[n] \\ &= \sum_{k=-\infty}^{\infty} f[k]g[n-k] \end{aligned} \tag{3.20}$$

As with continuous-time signals, the convolutional operator has the properties of commutative, linearity, and distribution for digital signals. That is, we can write

$$f[n] * g[n] = g[n] * f[n]$$
$$f[n] * (g[n] + k[n]) = f[n] * g[n] + f[n] * k[n]$$
$$f[n] * (ag[n] + bk[n]) = af[n] * g[n] + bf[n] * k[n]$$

Example 3.14 The graphs of $f[n]$ and $g[n]$ are given in Fig. 3.10. Find the convolution of these two signals.

Solution 3.14 Let us write the given signals in vector form as

$$f[n] = \begin{bmatrix} -1 & \underset{n=0}{\underbrace{0}} & 2 & 1 \end{bmatrix} \quad g[n] = \begin{bmatrix} 0 & \underset{n=0}{\underbrace{1}} & -1 \end{bmatrix}$$

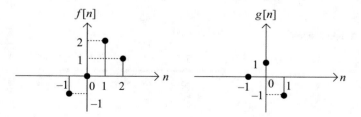

Fig. 3.10 Graphs of $f[n]$ and $g[n]$ for Example 3.14

The convolution of two digital signals can be calculated using the formula

$$h[n] = f[n] * g[n]$$

$$= \sum_{k=-\infty}^{\infty} f[k]g[n-k]$$

where giving values $-2, -1, 0, 1, 2$ for n, we get

$$h[-2] = \sum_{k=-\infty}^{\infty} f[k]g[-2-k] \quad h[-1] = \sum_{k=-\infty}^{\infty} f[k]g[-1-k]$$

$$h[0] = \sum_{k=-\infty}^{\infty} f[k]g[-k]$$

$$h[1] = \sum_{k=-\infty}^{\infty} f[k]g[1-k] \quad h[2] = \sum_{k=-\infty}^{\infty} f[k]g[2-k]$$

$$h[3] = \sum_{k=-\infty}^{\infty} f[k]g[3-k]$$

each of which can be calculated by expanding the summation term and inserting the signal values as

$$
\begin{aligned}
h[-2] &= \sum_{k=-\infty}^{\infty} f[k]g[-2-k] \\
&= \cdots + f[-2]g[0] + f[-1]g[-1] + f[0]g[-2] + \cdots \\
&= \cdots + 0 \times 0 + (-1) \times 0 + 0 \times 0 + \cdots \\
&= 0 \\
h[-1] &= \sum_{k=-\infty}^{\infty} f[k]g[-1-k] \\
&= \cdots + f[-1]g[0] + f[0]g[-1] + \cdots \\
&= \cdots + 1 \times (-1) + 0 + \cdots \\
&= -1
\end{aligned}
$$

resulting in

$$h[0] = 1 \quad h[1] = 2 \quad h[2] = -1 \quad h[3] = -1$$

Fig. 3.11 The graph of $h[n]$

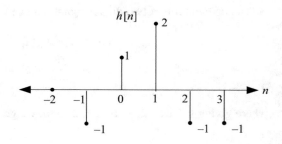

If

$$h[n] = \sum_{k=-\infty}^{\infty} f[k]g[n-k]$$

is calculated for n values other than $-2, -1, 0, 1, 2$, we obtain 0. The graph of $h[n]$ is depicted in Fig. 3.11.

Now let us see how to calculate the convolution using the visual method.

First, $f[k]$, $g[-k]$ signals are drawn. Then, $g[n-k]$ signals are obtained by shifting the $g[-k]$ signal for different n values. By finding the overlapping values of $g[n-k]$ and $f[k]$, and multiplying them and summing the multiplication results, $h[n]$ values are found.

The graphical solution is illustrated in Fig. 3.12. In this solution, we divide the horizontal axis of the $g[k]$ graph by -1 to get the graph of $g[-k]$. The graph of $g[n-k]$ for negative n values is obtained by shifting the graph of $g[-k]$ to the left by $|n|$ units, and in a similar manner, the graph of $g[n-k]$ for positive n values is obtained by shifting the graph of $g[-k]$ to the right by $|n|$ units.

Exercise Find the convolution of the digital signals

$$f[n] = \begin{bmatrix} \underset{n=0}{3.5} & -2.1 & 4.0 \end{bmatrix} \quad g[n] = \begin{bmatrix} 2 & \underset{n=0}{-1} & 3 \end{bmatrix}$$

Property Let the vector lengths of the digital signals $f[n]$ and $g[n]$ be N and M, respectively. The vector length of the digital signal $h[n]$ obtained from the convolution of these signals equals $N + M - 1$.

For continuous signals, the situation is somewhat different. Let the continuous-time signals $f(t)$ and $g(t)$ be defined on the intervals whose lengths are L_1 and L_2, respectively. The length of $h(t)$ obtained from the convolution of these two signals equals $L = L_1 + L_2$.

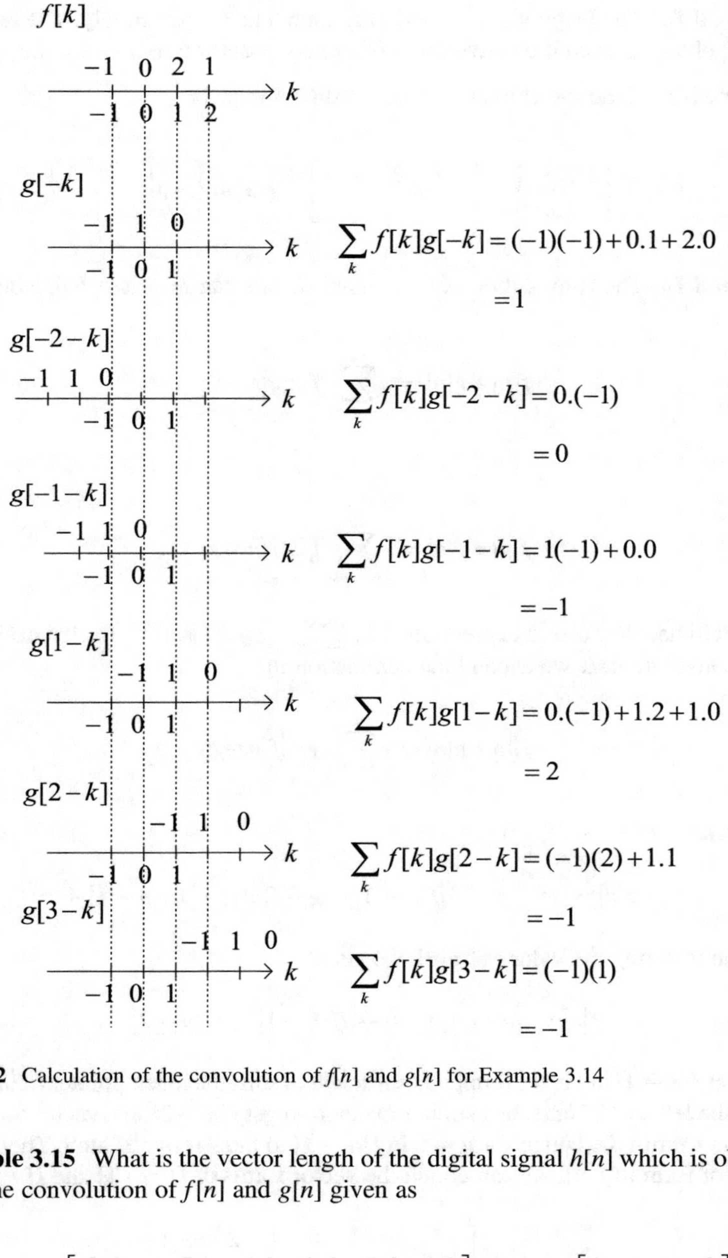

$f[k]$

$$\begin{array}{cccc} -1 & 0 & 2 & 1 \end{array}$$

$$\xrightarrow{} k$$

$$\begin{array}{cccc} -1 & 0 & 1 & 2 \end{array}$$

$g[-k]$

$$\begin{array}{ccc} -1 & 1 & 0 \end{array}$$

$$\xrightarrow{} k \qquad \sum_k f[k]g[-k] = (-1)(-1)+0.1+2.0$$

$$\begin{array}{ccc} -1 & 0 & 1 \end{array} \qquad\qquad = 1$$

$g[-2-k]$

$$\begin{array}{ccc} -1 & 1 & 0 \end{array}$$

$$\xrightarrow{} k \qquad \sum_k f[k]g[-2-k] = 0.(-1)$$

$$\begin{array}{ccc} -1 & 0 & 1 \end{array} \qquad\qquad = 0$$

$g[-1-k]$

$$\begin{array}{ccc} -1 & 1 & 0 \end{array}$$

$$\xrightarrow{} k \qquad \sum_k f[k]g[-1-k] = 1(-1)+0.0$$

$$\begin{array}{ccc} -1 & 0 & 1 \end{array} \qquad\qquad = -1$$

$g[1-k]$

$$\begin{array}{ccc} -1 & 1 & 0 \end{array}$$

$$\xrightarrow{} k \qquad \sum_k f[k]g[1-k] = 0.(-1)+1.2+1.0$$

$$\begin{array}{ccc} -1 & 0 & 1 \end{array} \qquad\qquad = 2$$

$g[2-k]$

$$\begin{array}{ccc} -1 & 1 & 0 \end{array}$$

$$\xrightarrow{} k \qquad \sum_k f[k]g[2-k] = (-1)(2)+1.1$$

$$\begin{array}{ccc} -1 & 0 & 1 \end{array} \qquad\qquad = -1$$

$g[3-k]$

$$\begin{array}{ccc} -1 & 1 & 0 \end{array}$$

$$\xrightarrow{} k \qquad \sum_k f[k]g[3-k] = (-1)(1)$$

$$\begin{array}{ccc} -1 & 0 & 1 \end{array} \qquad\qquad = -1$$

Fig. 3.12 Calculation of the convolution of $f[n]$ and $g[n]$ for Example 3.14

Example 3.15 What is the vector length of the digital signal $h[n]$ which is obtained from the convolution of $f[n]$ and $g[n]$ given as

$$f[n] = \begin{bmatrix} \underset{n=0}{3.5} & -2.1 & 4.0 & 2.4 & 5.6 & 1.2 \end{bmatrix} \quad g[n] = \begin{bmatrix} 2 & \underset{n=0}{-1} & 3 \end{bmatrix}$$

Solution 3.15 The lengths of $f[n]$ and $g[n]$ are 6 and 3, respectively. The length of the $h[n]$ obtained from the convolution of these signals is $6 + 3 - 1 = 8$.

Example 3.16 Find the convolution of the digital signals

$$f[n] = \begin{bmatrix} 1 & \underset{n=0}{-1} & 0 & 1 & 2 & -1 \end{bmatrix} \quad g[n] = \begin{bmatrix} \underset{n=0}{-1} & 2 & 1 \end{bmatrix}$$

Solution 3.16 The convolution of two digital signals can be calculated using either

$$f[n] * g[n] = \sum_{k=-\infty}^{\infty} f[k]g[n-k]$$

or

$$f[n] * g[n] = \sum_{k=-\infty}^{\infty} g[k]f[n-k]$$

We will use the second expression, i.e., $\sum_{k=-\infty}^{\infty} g[k]f[n-k]$, for the calculation of the convolution. If we expand the summation in

$$f[n] * g[n] = \sum_{k=-\infty}^{\infty} g[k]f[n-k]$$

we obtain

$$f[n] * g[n] = \cdots + g[-1]f[n+1] + g[0]f[n] + g[1]f[n-1] + \cdots$$

where substituting the values of $g[n]$, we get

$$f[n] * g[n] = -1f[n] + 2f[n-1] + 1f[n-2]$$

To get $f[n-1]$ in vector form, it is sufficient to shift the index pointer for $n = 0$ in $f[n]$ to the left by "1" unit. In a similar manner, to get $f[n-2]$ in a vector form, it is sufficient to shift the cursor for $n = 0$ in $f[n-1]$ to the left by "1" unit. Then, using the vector form of $f[n]$, we can obtain the vector forms of $f[n-1]$ and $f[n-2]$ as

$$f[n-1] = \begin{bmatrix} \underset{n=0}{1} & -1 & 0 & 1 & 2 & -1 \end{bmatrix}$$

$$f[n-2] = \begin{bmatrix} \underset{n=0}{0} & 1 & -1 & 0 & 1 & 2 & -1 \end{bmatrix}$$

Using the vector forms of $f[n-1]$ and $f[n-2]$ in

$$f[n] * g[n] = -1f[n] + 2f[n-1] + 1f[n-2]$$

we calculate the convolution as

$$f[n] * g[n] = -1 \times \begin{bmatrix} 1 & \underset{n=0}{-1} & 0 & 1 & 2 & -1 \end{bmatrix} + 2 \times \begin{bmatrix} \underset{n=0}{1} & -1 & 0 & 1 & 2 & -1 \end{bmatrix} + 1$$
$$\times \begin{bmatrix} \underset{n=0}{0} & 1 & -1 & 0 & 1 & 2 & -1 \end{bmatrix}$$

which is further simplified as

$$f[n] * g[n] = \begin{bmatrix} -1 & \underset{n=0}{1} & 0 & -1 & -2 & 1 \end{bmatrix} + \begin{bmatrix} \underset{n=0}{2} & -2 & 0 & 2 & 4 & -2 \end{bmatrix}$$
$$+ \begin{bmatrix} \underset{n=0}{0} & 1 & -1 & 0 & 1 & 2 & -1 \end{bmatrix}$$

where summing the elements of the vectors corresponding to the same indices, we obtain the convolution result as

$$f[n] * g[n] = \begin{bmatrix} -1 & \underset{n=0}{3} & -1 & -2 & 0 & 6 & 0 & -1 \end{bmatrix}$$

Exercise Find the convolution of the digital signals

$$x[n] = \begin{bmatrix} 1 & -1 & 1 & -1 \end{bmatrix} \quad y[n] = \begin{bmatrix} 1 & -1 & 1 & 1 \end{bmatrix}$$

3.2.5 Digital Impulse Signal (Function) and Convolution Operation

Digital impulse function is defined as

$$\delta[n] = \begin{cases} 1 & n=0 \\ 0 & \text{otherwise} \end{cases} \tag{3.21}$$

Any digital signal can be written as the sum of the shifted and scaled digital impulse signals, i.e.

$$f[n] = \ldots + a\delta[n-2] + b\delta[n-1] + c\delta[n] + d\delta[n+1] + e\delta[n+2] + \ldots$$

which can also be written using the summation symbol as

$$f[n] = \sum_{k=-\infty}^{\infty} f[k]\delta[n-k] \tag{3.22}$$

from which it is seen that any digital signal can be written as the convolution of itself and the digital impulse signal, that is:

$$f[n] = f[n] * \delta[n]$$

Some properties of the digital impulse signal are given as

$$f[n]\delta[n-n_0] = f[n_0]\delta[n-n_0]$$
$$f[n] * \delta[n-n_0] = f[n-n_0]$$
$$\sum f[n]\delta[n-n_0] = f[n_0]$$

Example 3.17 $\delta[n-1] * \delta[n-2] = ?$

Solution 3.17 Using the property $f[n] * \delta[n-n_0] = f[n-n_0]$ for the given convolution, we get

$$\delta[n-1] * \delta[n-2] = \delta[n-1-2]$$
$$= \delta[n-3]$$

3.3 Systems and Their Features

A system can be defined as an electronic circuit or any organization that receives a signal as input and produces another signal as output.

The system produces a new signal by processing the signal it receives as input and presents the processed signal as output. Inspecting the input and output of a system, we can get an idea about the properties of the system.

For example, a circuit containing a power supply, resistor, and capacitor has a memory, since a capacitor has energy storage property. But if our circuit consists of only the power supply and the resistor, there is no memory property. More complex systems can be obtained by connecting systems in series or parallel. A system can be analog or digital. In Fig. 3.13, systems are shown using block diagrams. A system can have the properties: causality, memory, linearity, time invariance, stability, and invertibility.

Fig. 3.13 Block diagram representation of systems

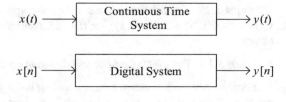

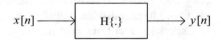

Fig. 3.14 A digital system

3.3.1 Mathematical Expressions of the Systems

The system takes the signal as input. Let us represent this signal as $x(t)$. It produces output by processing the input signal. Let us represent the output signal as $y(t)$. The relationship between output and input can be written as

$$y(t) = H\{x(t)\} \tag{3.23}$$

where $H\{\cdot\}$ represents the system operator and it is used to indicate that the system processes the input data.

If we have more than one system, we can use different letters to denote these systems, like $H_1\{\cdot\}$, $H_2\{\cdot\}$.

3.3.2 Causality

For a system to be causal, its output at any time must depend on current or past input data. If the output of the system depends on future input values, then the system is not causal.

Example 3.18 The relationship between the output and input of a system shown in Fig. 3.14 is given as $y[n] = x[n] + 0.5x[n - 2] + x^2[n - 3]$. Determine whether the system is causal or not.

Solution 3.18 The relationship between the output and the input of the system is $y[n] = x[n] + 0.5x[n - 2] + x^2[n - 3]$ from which it is seen that the output of the system $y[n]$ depends on the present input $x[n]$ and the past inputs $x[n - 2]$ and $x[n - 3]$.

$y[n]$ does not depend on any future input. The system, then, has causal property, or we can say that the system is causal.

Example 3.19 The relationship between the output and input of a system is given as $y(t) = x(t + 0.5) + x(t - 2.3) - x^3(t - 3)$. Determine whether the system is causal or not.

Solution 3.19 The system output $y(t)$ depends on the future time input $x(t + 0.5)$.

For this reason, the system is not causal, or we say that the system has no causality property. Systems without causality property can exist mathematically, but practically they cannot be constructed. Because non-causal systems need input data from the future to produce the output in the present, we cannot know the future input.

Especially real-time systems must have causality property. When researchers design systems, such as equalizers, they first design them mathematically. Then, they try to make the system causal. This is necessary to get practical systems; otherwise, the designed systems will remain only theoretical and have no practical value.

3.3.3 Memory

If the output of a system depends only on the current input, then the system has no memory.

If the output of the system depends on past or future input data, then the system has memory, or the system has memory property.

Example 3.20 The relationship between the output and the input of a system is given by $y[n] = x[n] + x^3[n]$. Determine whether the system has memory or not.

Solution 3.20 Since the output of the system $y[n]$ at time n is only a function of the current input data $x[n]$, the system has no memory.

Example 3.21 The relationship between the output and the input of a system is given by $y(t) = x(t - 1) + x^2(t)$. Determine whether the system has memory or not.

Solution 3.21 Since the output of the system $y(t)$ at the time instant t depends on past input $x(t - 1)$, the system has memory.

3.3.4 Stability

Systems can have bounded-input, bounded-output stability property. If we get bounded output when we give bounded-input data to a system, then the system is stable. The system on the other hand is unstable if for the bounded input the output is unbounded. Let us give two examples to understand the subject. Consider an iron jerry can and have two small holes on the faces of this jerry can. Assume that the jerry can is filled with gasoline; if we throw a burning match into one of the holes and

Fig. 3.15 Illustration of
stable and unstable systems

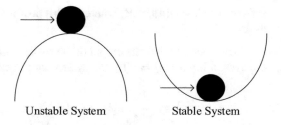

Unstable System   Stable System

close it, a high amount of pressure, heat, and light will appear at the other hole; and
then we have an unstable system.

On the other hand, consider a circuit consisting of only a voltage source and a
resistor connected in parallel. The more you increase the source voltage, the more
voltage will be obtained across the resistor, so our system is a stable system. The
magnitude of the voltage across the resistor will be proportional to the magnitude of
the source voltage.

If the input and output of a system are denoted by $x(t)$ and $y(t)$, then for a
bounded-input and bounded-output stable system, for

$$|x(t)| \leq M_x < \infty$$

we have

$$|y(t)| \leq M_y < \infty$$

where M_x and M_y are real numbers. Unless otherwise indicated we will use the word
"stability" for "bounded-input, bounded-output stability," and if it is mentioned that
a system is stable, it means that we indicate a bounded-input, bounded-output stable
system. In Fig. 3.15, graphical illustration of stable and unstable systems is available.
As can be interpreted from Fig. 3.15, an unstable system produces an immense
amount of output for very little input, while a stable system produces a small amount
of output for small input.

Example 3.22 The relationship between the output and the input of a system is
given by

$$y[n] = \sum_{k=-\infty}^{n} x[k]$$

Determine whether the system is stable or not.

Solution 3.22 To determine whether the system is stable or not, let us consider
bounded-input data to the system, and see if the output of the system is bounded as
well. For the system input, let us use $x[n] = u[n]$ which is unit step function that

satisfies $|x[n]| = |u[n]| < M_x$ where M_x is a positive real number and it can be chosen as $M_x = 2$.

Now let us calculate the output of the system and see if the output is bounded. For the system input $x[n] = u[n]$, the system output can be calculated as

$$y[n] = \sum_{k=-\infty}^{n} u[k] \rightarrow y[n] = \sum_{k=0}^{n} 1 \rightarrow y[n] = n + 1, \quad n > 0$$

For the system output $y[n] = n + 1$, it is not possible to find a real number M_y such that $|y[n]| < M_y$. Since n goes to infinity, the output $y[n] = n + 1$ also goes to infinity, and we cannot find an upper bound for the output of the system.

Thus, the system is an unstable system; it does not have stability property.

Exercise The relationship between the output and the input of a system is given as

$$y(t) = x^2(t) + x(t - 3)$$

Determine whether the system is stable or not.

3.3.5   Invertibility

If the input of a system can be obtained from its output, then the system is invertible. In other words, a system is invertible if it is possible to have another system from which the input of the system can be obtained using its output. Consider a system where the relationship between the output and the input of the system is given by $y[n] = x[n - 1]$, and let us denote this system by H_1.

For a second system, the relationship between the output and the input of the system is given by $k[n] = m[n + 1]$, and let us denote this system by H_2. If $y[n]$ is taken as input by H_2, then the output of H_2 can be calculated as

$$
\begin{aligned}
k[n] &= m[n + 1] \\
&= y[n + 1] \\
&= x[n - 1 + 1] \\
&= x[n]
\end{aligned}
$$

which is nothing but the input of H_1. Hence, H_2 is the inverse of H_1.

3.3.6 Linearity

Linearity is another property that a system can have. Let x_1 and x_2 be two different inputs of the system. Let the output of the system for these inputs be y_1 and y_2. For the system to be a linear system, for the input $k_1 x_1 + k_2 x_2$ $k_1, k_2 \in R$, the output must be $k_1 y_1 + k_2 y_2$.

In order to understand whether a system is linear or not, we only need to look at the equation between its output and input. If the output of the system includes second or higher powers of the input, then the system is not linear.

Example 3.23 Consider a system with an input-output relationship

$$y[n] = x^2[n-1]$$

determines whether the system is linear or not.

Solution 3.23 Since the output of the system depends on the square of the input signal, we can directly say that the system is not linear.

Applying the definition of linearity, let us see mathematically that it is not linear. For inputs $x_1[n]$ and $x_2[n]$, the outputs are $y_1[n] = x_1^2[n-1]$ and $y_2[n] = x_2^2[n-1]$. If the system output for the input $x_1[n] + x_2[n]$, $k_1 = k_2 = 1$ equals $y_1[n] + y_2[n]$, then the system is linear. For input $x_1[n] + x_2[n]$, the output is $(x_1[n-1] + x_2[n-1])^2$ which is not equal to $y_1[n] + y_2[n]$. Hence, the system is not linear.

Example 3.24 The relationship between input and output of a system is given as $y(t) = x(t)x(t-1)$. Determine whether the system is linear or not.

Solution 3.24 For input $x_1(t)$, the output is $y_1(t) = x_1(t)x_1(t-1)$. For input $x_2(t)$, the output is $y_2(t) = x_2(t)x_2(t-1)$. For input $x_1(t) + x_2(t)$, the output is

$$
\begin{aligned}
y(t) &= (x_1(t) + x_2(t))(x_1(t-1) + x_2(t-1)) \\
&= x_1(t)x_1(t-1) + x_1(t)x_2(t-1) + x_2(t)x_1(t-1) + x_2(t)x_2(t-1)
\end{aligned}
$$

which is not equal to $y_1(t) + y_2(t)$, i.e., $y(t) \neq y_1(t) + y_2(t)$. Hence, the system is not linear.

Exercise Determine whether the system defined by $y[n] = x[n] + 2x[n-1]$ is linear or not.

3.3.7 Time Invariance

For a system to be time-invariant, the time-shifted input must produce time-shifted output. The shift amount at the input and output must be the same.

That is, for the input signal $x(t)$ if the output is $y(t)$, then for a time-invariant system, for the input $x(t - t_0)$, the output is $y(t - t_0)$, or for the input $x(t + t_0)$, the output is $y(t + t_0)$.

Example 3.25 The relationship between input and output of a system is given by

$$y(t) = \frac{x(t)}{R(t)}$$

Determine whether the system is time-invariant or not.

Solution 3.25 For input $x(t)$, the output is

$$\frac{x(t)}{R(t)}$$

For input $x(t - t_0)$, the output is

$$\frac{x(t - t_0)}{R(t)}$$

For the system to be time-invariant, for the input $x(t - t_0)$, the output must be

$$y(t - t_0) = \frac{x(t - t_0)}{R(t - t_0)}$$

However,

$$\frac{x(t - t_0)}{R(t)} \neq y(t - t_0) = \frac{x(t - t_0)}{R(t - t_0)}$$

Hence, the system is time-dependent, and it is not time-invariant.

Example 3.26 Show that the system defined by

$$y(t) = \int_{-\infty}^{t} x(\tau)d\tau$$

is time-invariant.

Solution 3.26 For input $x(t - t_0)$, the output is

$$y_1(t) = \int\limits_{-\infty}^{t} x(\tau - t_0) d\tau$$

For the time-invariant system, we have

$$y_1(t) = y(t - t_0)$$

Making parameter change in

$$y(t - t_0) = \int\limits_{-\infty}^{t - t_0} x(\tau) d\tau$$

such that $\tau' = \tau + t_0$ we get

$$y(t - t_0) = \int\limits_{-\infty}^{t} x(\tau') d\tau'$$

from which we can write that

$$y(t - t_0) = y_1(t)$$

Hence, the system is time-invariant.

Exercise Consider a system with an input-output relationship $y(t) = x(2t)$. Determine whether the system is time-invariant or not.

Exercise Consider a system with an input-output relationship $y(t) = x(t) + u(t)$. Determine whether the system is time-invariant or not.

Exercise Consider a system with an input-output relationship $y(t) = x(t - 1) + u(2t)$. Determine whether the system is time-invariant or not.

3.3.8 Time Domain Representations of Linear and Time-Invariant Systems

Any digital signal $x[n]$ can be written as the sum of the shifted impulse functions. The digital signal in Fig. 3.16 can be written as

$$x[n] = -\delta[n + 2] + 2\delta[n + 1] + \delta[n - 2] - 2\delta[n - 4]$$

In general the digital signal $x[n]$ can be written as

Fig. 3.16 A digital signal

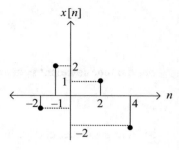

Fig. 3.17 A digital system

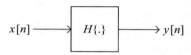

Fig. 3.18 Impulse response
of an LTI system

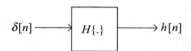

$$x[n] = \sum_{k=-\infty}^{\infty} x[k]\delta[n-k] \tag{3.24}$$

where $x[k]$ is the amplitude at time instant k. The equation (3.24) can also be written
as

$$x[n] = x[n] * \delta[n] \tag{3.25}$$

Consider a system $H\{\cdot\}$ having linearity and time-invariance properties. Such a
system can also be called a linear and time-invariant system, i.e., LTI system, in
short. Assume that the LTI system has input $x[n]$ and output $y[n]$ as depicted in
Fig. 3.17.

For input $x[n] = \delta[n]$, let us denote the output by $y[n] = h[n]$ as illustrated in
Fig. 3.18.

In Fig. 3.17, the relationship between $y[n]$ and $x[n]$ can be expressed as

$$y[n] = H\{x[n]\}$$

where substituting

$$x[n] = \sum_{k=-\infty}^{\infty} x[k]\delta[n-k]$$

we get

$$y[n] = H\{x[n]\}$$
$$= H\left\{ \sum_{k=-\infty}^{\infty} x[k]\delta[n-k] \right\}$$

where moving $H\{\cdot\}$ operator inside the summation term we obtain

$$y[n] = \sum_{k=-\infty}^{\infty} x[k]H\{\delta[n-k]\} \qquad (3.26)$$

The system is an LTI system, and for

$$x[n] = \delta[n]$$

the output is

$$y[n] = h[n]$$

Due to the time-invariance property, for input

$$x[n] = \delta[n-k]$$

the system output happens to be

$$y[n] = h[n-k]$$

that is, if

$$h[n] = H\{\delta[n]\}$$

then we have

$$h[n-k] = H\{\delta[n-k]\}$$

When

$$h[n-k] = H\{\delta[n-k]\}$$

is substituted in (3.26), we get

$$y[n] = \sum_{k=-\infty}^{\infty} x[k]h[n-k]$$

which can be also written as

$$y[n] = x[n] * h[n] \tag{3.27}$$

which means that if the output of the system for the impulse input is known, the output for any input can be calculated by taking the convolution of input and impulse response.

The special output $h[n]$ obtained for impulse input is called impulse response.

Note that to use

$$y[n] = x[n] * h[n]$$

property, the system should have linearity and time-invariance properties, i.e., it should be an LTI system,

If the system is not an LTI system, then we can still calculate the impulse response of the system; however, we cannot use the equation

$$y[n] = x[n] * h[n]$$

to calculate the output for an arbitrary input signal. Although we explained the subject for digital systems, the results we obtained are also valid for continuous-time systems. Let $h(t)$ be the impulse response of a continuous LTI system. The system output for input $x(t)$ can be calculated using

$$y(t) = x(t) * h(t) \tag{3.28}$$

Example 3.27 Consider an LTI system with an input-output relationship $y[n] = x[n] + 0.5x[n - 2]$. Find the impulse response of the system.

Solution 3.27 The impulse response of the system is the output of the system when the input data is impulse.

For input $x[n] = \delta[n]$, the system output can be calculated as

$$h[n] = \delta[n] + 0.5\delta[n - 2]$$

which can also be written as

$$h[n] = \begin{cases} 1 & n = 0 \\ 0.5 & n = 2 \\ 0 & \text{otherwise} \end{cases}$$

Example 3.28 If $h[n]$ is the impulse response of an LTI system, find the output of the system in terms of $h[n]$ for the input

Fig. 3.19 Impulse response
for a digital system

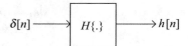

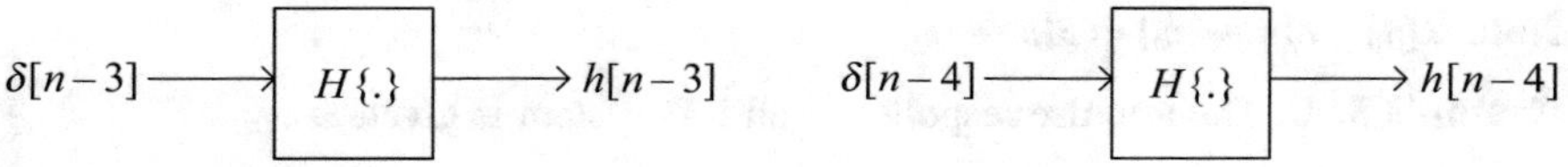

$$\delta[n-3] \longrightarrow \boxed{H\{.\}} \longrightarrow h[n-3] \qquad \delta[n-4] \longrightarrow \boxed{H\{.\}} \longrightarrow h[n-4]$$

Fig. 3.20 Shifted impulse responses for a digital system

$$x[n] = \begin{cases} 1 & n=0 \\ 2 & n=3 \\ -0.5 & n=4 \end{cases}$$

Solution 3.28 Let us express the input signal in terms of the impulse function as in

$$x[n] = \delta[n] + 2\delta[n-3] - 0.5\delta[n-4]$$

When the input signal is impulse, the output of the system is $h[n]$ as illustrated in Fig. 3.19.

Since the system has time-invariance property, then for the input $\delta[n-3]$, the output is $h[n-3]$, and for the input $\delta[n-4]$, the output is $h[n-4]$ as shown in Fig. 3.20.

Since the system has also linearity property, then for the input signal

$$x[n] = \delta[n] + 2\delta[n-3] - 0.5\delta[n-4]$$

the system output happens to be

$$y[n] = h[n] + 2h[n-3] - 0.5h[n-4]$$

Example 3.29 The impulse response of an LTI system is given as

$$h[n] = \delta[n] - \delta[n-4]$$

Calculate the output of the system for input $x[n] = u[n]$.

Solution 3.29 If impulse response $h[n]$ of an LTI system is known, then the system output for an arbitrary input $x[n]$ can be calculated as

$$y[n] = h[n] * x[n]$$

where substituting $h[n] = \delta[n] - \delta[n-4]$, we obtain

$$y[n] = (\delta[n] - \delta[n-4]) * u[n]$$
$$= \delta[n] * u[n] - \delta[n-4] * u[n]$$
$$= u[n] - u[n-4]$$

Note $x[n] * \delta[n - n_0] = x[n - n_0]$

Example 3.30 The impulse response of an LTI system is given as

$$h[n] = \delta[n] + \frac{1}{2}\delta[n-1]$$

Calculate the output of the system for the input

$$x[n] = \begin{cases} 2 & n=0 \\ 4 & n=1 \\ 0 & \text{otherwise} \end{cases}$$

Solution 3.30-(1) The input signal can be expressed in terms of impulse functions as

$$x[n] = 2\delta[n] + 4\delta[n-1]$$

We can calculate the system output using

$$y[n] = h[n] * x[n]$$

where substituting

$$x[n] = 2\delta[n] + 4\delta[n-1] \quad h[n] = \delta[n] + \frac{1}{2}\delta[n-1]$$

we obtain

$$y[n] = \left(\delta[n] + \frac{1}{2}\delta[n-1]\right) * (2\delta[n] + 4\delta[n-1])$$
$$= 2\delta[n] * \delta[n] + 4\delta[n] * \delta[n-1] + \delta[n-1] * \delta[n] + 2\delta[n-1] * \delta[n-1]$$
$$= 2\delta[n] + 4\delta[n-1] + \delta[n-1] + 2\delta[n-2]$$
$$= 2\delta[n] + 5\delta[n-1] + 2\delta[n-2]$$

Solution 3.30-(2) Using the impulse response for the given input, the system output using the linearity and time-invariance property can be written as

$$y[n] = 2h[n] + 4h[n-1]$$

in which substituting

$$h[n] = \delta[n] + \frac{1}{2}\delta[n-1]$$

we obtain

$$y[n] = 2\left(\delta[n] + \frac{1}{2}\delta[n-1]\right) + 4\left(\delta[n-1] + \frac{1}{2}2\delta[n-2]\right)$$
$$= 2\delta[n] + \delta[n-1] + 4\delta[n-1] + 2\delta[n-2]$$
$$= 2\delta[n] + 5\delta[n-1] + 2\delta[n-2]$$

Example 3.31 The impulse response of an LTI system is given as

$$h[n] = \left(\frac{3}{4}\right)^n u[n]$$

Calculate the step response of the system.

Solution 3.31 Step response is the output of the system when the input is a step function. For input $x[n] = u[n]$, the system output can be calculated as

$$y[n] = x[n] * h[n]$$
$$= u[n] * \left(\frac{3}{4}\right)^n u[n]$$
$$= \sum_{k=-\infty}^{\infty} u[k]\left(\frac{3}{4}\right)^{n-k} u[n-k]$$
$$= \sum_{k=0}^{n} \left(\frac{3}{4}\right)^{n-k}$$
$$= \left(\frac{3}{4}\right)^n \sum_{k=0}^{n} \left(\frac{3}{4}\right)^{-k}$$
$$= \left(\frac{3}{4}\right)^n \sum_{k=0}^{n} \left(\frac{4}{3}\right)^{k}$$
$$= \frac{\left(\frac{3}{4}\right)^n \left(1 - \left(\frac{4}{3}\right)^{n+1}\right)}{1 - \frac{4}{3}}$$
$$= -3\left(\frac{3}{4}\right)^n \left(1 - \left(\frac{4}{3}\right)^{n+1}\right)$$

Hence, the step response of the system is

$$y[n] = -3\left(\frac{3}{4}\right)^n \left(1 - \left(\frac{4}{3}\right)^{n+1}\right)$$

Note We can use $s[n]$ to indicate the step response of a system.

Example 3.32 The impulse response of a continuous-time system is given as $h(t) = \delta(t) - \delta(t - 2)$. Calculate the output of the system for the input $x(t) = \left(\frac{1}{2}\right)^t u(t)$.

Solution 3.32

$$y(t) = x(t) * h(t)$$

$$= \left(\left(\frac{1}{2}\right)^t u(t)\right) * (\delta(t) - \delta(t - 2))$$

$$= \left(\frac{1}{2}\right)^t u(t) - \left(\frac{1}{2}\right)^{t-2} u(t - 2)$$

3.3.9   *The Relationship Between Impulse and Unit Step Responses of a Linear and Time-Invariant System*

The relationships between impulse response $h[n]$ and step response $s[n]$ of a digital system are given as

$$h[n] = s[n] - s[n - 1]$$

$$s[n] = \sum_{k=-\infty}^{n} h[k] \tag{3.29}$$

For a continuous-time system, the relationships between $h(t)$ and $s(t)$ are given as

$$h(t) = \frac{ds(t)}{dt}$$

$$s(t) = \int_{\tau=-\infty}^{t} h(\tau)dt \tag{3.30}$$

3.3.10   The Response of a Linear and Time-Invariant System to an Exponential Signal

Let $h[n]$ be the impulse response of an LTI digital system. For exponential input $x[n] = e^{j\Omega n}$, the system output is calculated as

$$y[n] = x[n] * h[n]$$

$$= \sum_{k=-\infty}^{\infty} h[k]x[n-k]$$

$$= \sum_{k=-\infty}^{\infty} h[k]e^{j\Omega(n-k)}$$

$$= e^{j\Omega n} \sum_{k=-\infty}^{\infty} h[k]e^{-j\Omega k}$$

where the expression

$$\sum_{k=-\infty}^{\infty} h[k]e^{-j\Omega k}$$

is the Fourier transform of $h[n]$, and it is indicated by $H(\Omega)$. Hence, the system output for the exponential input signal can be written as

$$y[n] = e^{j\Omega n} H(\Omega) \qquad (3.31)$$

Some authors define the Fourier transform as

$$H(\Omega) = \frac{1}{\sqrt{2\pi}} \sum_{k=-\infty}^{\infty} h[k]e^{-j\Omega k}$$

In this case, output is written as

$$y[n] = e^{j\Omega n} \sqrt{2\pi} H(\Omega)$$

Now, let us try to calculate the output of a linear and time-invariant system for cosine input signal. The input signal

$$x[n] = \cos(\Omega n + \psi)$$

can be written in terms of exponential signals as

$$x[n] = \cos\left(\Omega n + \psi\right)$$
$$= \frac{1}{2}\left(e^{j(\Omega n + \psi)} + e^{-j(\Omega n + \psi)}\right)$$
$$= \frac{1}{2}\left(e^{j\Omega n}e^{j\psi} + e^{-j\Omega n}e^{-j\psi}\right)$$

Considering the result in (3.31), the system output for the cosine input signal can be written as

$$y[n] = \frac{1}{2}\left(H(\Omega)e^{j\Omega n}e^{j\psi} + H^*(\Omega)e^{-j\Omega n}e^{-j\psi}\right) \tag{3.32}$$

where $H^*(\Omega)$ is the complex conjugate of $H(\Omega)$.

The complex function $H(\Omega)$ can be written as

$$H(\Omega) = |H(\Omega)| \angle H(\Omega) \tag{3.33}$$

When (3.33) is substituted into (3.32), we obtain

$$y[n] = \frac{1}{2}\left(|H(\Omega)|e^{j\angle|H(\Omega)|}e^{j\Omega n}e^{j\psi} + |H(\Omega)|e^{-j\angle|H(\Omega)|}e^{-j\Omega n}e^{-j\psi}\right)$$

which can be written in a more compact form as

$$y[n] = |H(\Omega)| \cos\left(\angle H(\Omega) + \Omega n + \psi\right) \tag{3.34}$$

Note $e^{j\theta} = \cos\theta + j\sin\theta \quad \sin\theta = \frac{1}{2j}\left(e^{j\theta} - e^{-j\theta}\right) \quad \cos\theta = \frac{1}{2}\left(e^{j\theta} + e^{-j\theta}\right)$

Example 3.33 The impulse response of an LTI system is given as

$$h[n] = \delta[n-1] + \delta[n+1]$$

For input $x[n] = \cos\left(\Omega n + \frac{\pi}{3}\right)$, calculate the system output.

Solution 3.33 We can use the formula

$$y[n] = |H(\Omega)| \cos\left(\angle H(\Omega) + \Omega n + \psi\right) \tag{3.35}$$

for the output calculation. The Fourier transform expression in (3.35) can be calculated as

$$H(\Omega) = \sum_{k=-\infty}^{\infty} h[k] e^{-j\Omega k}$$

$$= \sum_{k=-\infty}^{\infty} (\delta[k-1] + \delta[k+1]) e^{-j\Omega k} \qquad (3.36)$$

$$= e^{-j\Omega} + e^{j\Omega}$$

$$= 2\cos(\Omega)$$

Substituting (3.36) in (3.35), we obtain

$$y[n] = 2|\cos(\Omega)| \cos\left(\Omega n + \frac{\pi}{3} + \angle \cos(\Omega)\right)$$

where

$$\angle \cos(\Omega) = \begin{cases} 2m\pi & k2\pi \le \Omega < \frac{\pi}{2} + k2\pi \text{ or } \frac{3\pi}{2} + k2\pi \le \Omega < (k+1)2\pi \\ \pi + 2m\pi & \text{else} \end{cases}$$

where $k, m \in Z$.

Example 3.34 The impulse response of an LTI system is given as

$$h[n] = \delta[n-2]$$

For input $x[n] = \sin\left(\Omega n + \frac{\pi}{4}\right)$, calculate the system output $y[n]$.

Solution 3.34 The input signal is a sine signal which is a phase-shifted version of a cosine signal, and it can be written in terms of a cosine signal, and we can use the expression

$$y[n] = |H(\Omega)| \cos(\angle H(\Omega) + \Omega n + \psi)$$

to calculate the system output. We leave the rest of the solution to the reader.

Note Although we considered only digital systems up to this point, similar formulas can be derived for continuous-time systems as well. If the impulse response of a continuous LTI system is known, then the output of the continuous-time system for the exponential input e^{jwt} can be calculated as

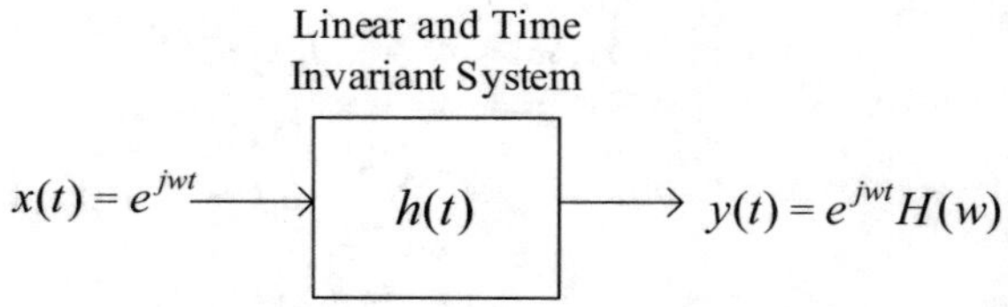

Fig. 3.21 The response of
an LTI continuous system to
an exponential input

Linear and Time
Invariant System

$$x(t) = e^{jwt} \longrightarrow \boxed{h(t)} \longrightarrow y(t) = e^{jwt}H(w)$$

$$H(w) = \int\limits_{\tau=-\infty}^{\infty} h(\tau)e^{-jw\tau}\,d\tau$$

$$y(t) = x(t) * h(t)$$

$$= \int\limits_{\tau=-\infty}^{\infty} h(\tau)x(t-\tau)d\tau$$

$$= \int\limits_{\tau=-\infty}^{\infty} h(\tau)e^{jw(t-\tau)}d\tau$$

$$= e^{jwt}\int\limits_{\tau=-\infty}^{\infty} h(\tau)e^{-jwt}d\tau$$

$$= e^{jwt}H(w)$$

where $H(w)$ is the Fourier transform of the impulse function. In Fig. 3.21, a
continuous-time system with its exponential input and output is shown.

3.3.11   Determining the Properties of a Linear and Time-Invariant System from Impulse Response

The properties of systems are memory, causality, linearity, time invariance, and
stability. If a system is linear and time-invariant, the rest of its properties (stability,
causality, memory) can be determined directly from the system's impulse response.

Memory

If the impulse responses of linear and time-invariant systems (digital or continuous-
time) are in the form

$$h[n] = K\delta[n] \quad h(t) = M\delta(t) \tag{3.37}$$

then the systems are memoryless; otherwise, the systems have memory.

Proof The output of a digital LTI system can be calculated using

$$y[n] = \sum_k h[k]x[n-k]$$

where the output of the system $y[n]$ does not depend on the past and future input data if we have

$$h[k] = \begin{cases} K & k=0 \\ 0 & \text{otherwise} \end{cases}$$

Otherwise, when

$$y[n] = \sum_k h[k]x[x-k]$$

is expanded, $y[n]$ will contain input data from the past or future. This means that the system has memory. For continuous-time systems, a similar proof can be achieved.

Causality

For an LTI system to be a causal system, its impulse response must satisfy, depending on whether the system is digital or continuous, one of the expressions

$$h[n] = 0 \quad n<0, \quad h(t) = 0 \quad t \le 0 \tag{3.38}$$

Proof System output can be expressed as

$$y[n] = \sum_k h[k]x[n-k]$$

which depends on future input values, i.e., $x[n+1]$, $x[n+2]$, ..., for negative k values. If

$$h[k] = 0 \quad k<0$$

then $y[n]$ does not contain future inputs, and we have a causal system.

Stability

For an LTI system to be a stable system, its impulse response must satisfy, depending on whether the system is digital or continuous, one of the expressions

$$\sum_k |h[k]| < \infty \qquad \int_{-\infty}^{\infty} |h(\tau)| d\tau < \infty \qquad (3.39)$$

Proof The system output can be written as

$$y[n] = \sum_k h[k]x[n-k]$$

where taking the absolute values of both sides and using $|A + B| < |A| + |B|$ and $|AB| = |A||B|$ we get

$$|y[n]| = \left| \sum_k h[k]x[n-k] \right|$$
$$\leq \sum_k |h[k]||x[n-k]|$$

where using $|x[n-k]| \leq N$, since the input is bounded, we obtain

$$|y[n]| \leq \sum_k N|h[k]|$$

where the right hand side is a finite number if we have

$$\sum_k |h[k]| < \infty$$

Hence, for a digital LTI system to be a stable system, its impulse response should be absolutely summable.

Example 3.35 An LTI system is depicted in Fig. 3.22. Determine whether the system is stable or not.

Solution 3.35 For a digital LTI system to be a stable system, its impulse response should be absolutely summable, i.e., impulse response must satisfy

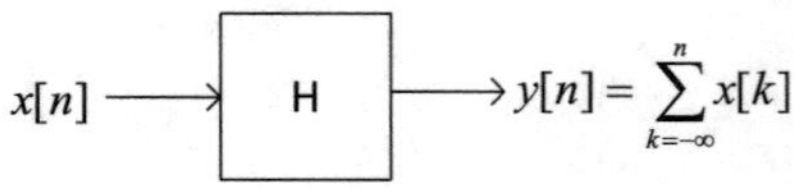

Fig. 3.22 An LTI system

$$x[n] \longrightarrow \boxed{\text{H}} \longrightarrow y[n] = \sum_{k=-\infty}^{n} x[k]$$

$$\sum_k |h[k]| < \infty$$

For the system depicted in Fig. 3.22, the impulse response can be calculated as

$$h[n] = \sum_{k=-\infty}^{n} \delta[k]$$
$$= u[n]$$

for which if we use (3.39), we obtain

$$\sum_k |u[k]| = \sum_{k=0}^{n} 1$$
$$= \infty$$

which is not a finite number. Since the absolute sum of the impulse response is not a finite number, then the system is an unstable system.

Example 3.36 The input-output relationship of a digital system is shown in Fig. 3.23. Find the impulse response of this system.

Solution 3.36 For $x[n] = \delta[n]$ the output is denoted by $y[n] = h[n]$. For impulse input, the output of the system can be written as

$$h[n] = \rho h[n-1] + \delta[n] \tag{3.40}$$

Assuming that $h[n] = 0$, $n < 0$, (3.40) can be recursively calculated as

n	$h[n]$
0	1
1	ρ
2	ρ^2
3	ρ^3
.	.
.	.

from which we can write the mathematical expression of the impulse response as

$$h[n] = \rho^n u[n]$$

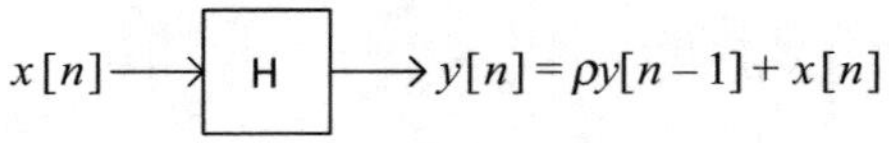

Fig. 3.23 The input-output relationship of a digital system

$$x[n] \longrightarrow \boxed{H} \longrightarrow y[n] = \rho y[n-1] + x[n]$$

Example 3.37 Input-output relationship of a system is given as

$$y[n] = x[Mn], \quad M \in Z$$

Determine whether the system has time-invariance property or not.

Solution 3.37 For input

$$x[n - n_0]$$

the system output is

$$y_1[n] = x[Mn - n_0]$$

We have

$$y[n - n_0] = x[M(n - n_0)]$$

Since

$$y_1[n] \neq y[n - n_0]$$

the system does not have time-invariance property.

3.3.12   Representation of Linear and Time-Invariant Systems with Differential and Difference Equations

The output of a system can be expressed not only in terms of system input but also in terms of both system input and output.

These equations need to be solved in order to express the instantaneous output of the system in terms of system inputs.

Continuous-Time Systems

A continuous LTI system can be expressed by the differential equation

$$\sum_{k=0}^{N} a_k \frac{d^k y(t)}{dt^k} = \sum_{k=0}^{M} b_k \frac{d^k x(t)}{dt^k} \tag{3.41}$$

Digital Systems

A digital LTI system can be expressed by the difference equation

$$\sum_{k=0}^{N} a_k y[n-k] = \sum_{k=0}^{M} b_k x[n-k] \tag{3.42}$$

3.3.13   Solving Differential and Difference Equations of Linear and Time-Invariant Systems

In this section, we will learn how to solve differential and difference equations of systems. First, let us see how to solve the differential equations used for continuous-time systems.

Differential Equation Solution

The solution of the differential equation

$$\sum_{k=0}^{N} a_k \frac{d^k y(t)}{dt^k} = \sum_{k=0}^{M} b_k \frac{d^k x(t)}{dt^k} \tag{3.43}$$

contains two parts. The first part is called the homogeneous solution, and the second part is called the particular solution.

Homogeneous solution is obtained from the equation

$$\sum_{k=0}^{N} a_k \frac{d^k y(t)}{dt^k} = 0 \tag{3.44}$$

whose solution is in the form

$$y_h(t) = \sum_{k=1}^{N} c_i e^{r_i t} \tag{3.45}$$

where r_i satisfy

$$\sum_{k=0}^{N} a_k r^k = 0 \tag{3.46}$$

and the coefficients c_i are determined from the initial conditions. The equation

$$\sum_{k=0}^{N} a_k r^k = 0 \tag{3.47}$$

is called characteristic equation.

Example 3.38 Find the homogeneous solution, $y_h(t)$, of

$$\frac{d^2 y(t)}{dt^2} + 5\frac{dy(t)}{dt} + 6y(t) = 2x(t) + \frac{dx(t)}{dt}$$

Solution 3.38 To find the homogeneous solution, let us consider the equation

$$\frac{d^2 y(t)}{dt^2} + 5\frac{dy(t)}{dt} + 6y(t) = 0$$

whose roots satisfy

$$\sum_{k=0}^{N} a_k r^k = 0$$

from which we get

$$r^2 + 5r + 6 = 0$$

whose roots are $r = -3, -2$. Then, homogeneous solution can be written as

$$y_h(t) = c_1 e^{-3t} + c_2 e^{-2t}$$

Let us now consider the particular solution of

$$\sum_{k=0}^{N} a_k \frac{d^k y(t)}{dt^k} = \sum_{k=0}^{M} b_k \frac{d^k x(t)}{dt^k} \tag{3.48}$$

Particular solution is the solution obtained for certain input signals. Considering the input signal $x(t)$, particular solution of (3.48) can be in one of the forms shown in Table 3.1.

Table 3.1 Solution table for continuous-time differential equations

Input $x(t)$	Particular solution $y_p(t)$
K (constant number)	C (constant number)
t	$c_1 t + c_2$
e^{-at}	ce^{-at}
t^n	$c_1 t^n + c_2 t^{n-1} + \ldots + c_n$
$\cos(wt + \phi)$	$c_1 \cos(wt) + c_2 \sin(wt)$

The total solution of

$$\sum_{k=0}^{N} a_k \frac{d^k y(t)}{dt^k} = \sum_{k=0}^{M} b_k \frac{d^k x(t)}{dt^k}$$

is obtained summing homogeneous and particular solutions, i.e.,

$$y(t) = y_h(t) + y_p(t)$$

Example 3.39 For input $x(t) = e^{-t}$, find the particular solution of

$$\frac{d^2 y(t)}{dt^2} + 5\frac{dy(t)}{dt} + 6y(t) = 2x(t) + \frac{dx(t)}{dt} \tag{3.49}$$

Solution 3.39 For the given input, it is seen from Table 3.1 that the particular solution is in the form

$$y_p(t) = Ke^{-t} \tag{3.50}$$

If we substitute the input and particular solution in (3.49), we get

$$Ke^{-t} - 5Ke^{-t} + 6Ke^{-t} = 2e^{-t} - e^{-t}$$

from which the coefficient K is found as $K = \frac{1}{2}$. Substituting the calculated coefficient in (3.50), we obtain the particular solution as

$$y_p(t) = \frac{1}{2}e^{-t}$$

Example 3.40 Find the total solution of

$$\frac{d^2y(t)}{dt^2} + 5\frac{dy(t)}{dt} + 6y(t) = 2x(t) + \frac{dx(t)}{dt}$$

for the input $x(t) = e^{-t}$.

Solution 3.40 In the previous two examples, we considered the same equation and found homogeneous and particular solutions. Summing these two solutions, we obtain the total solution as

$$y(t) = y_h(t) + y_p(t)$$
$$= c_1 e^{-3t} + c_2 e^{-2t} + \frac{1}{2}e^{-t}$$

Solving Difference Equations

The solution of the difference equations includes two solutions as in the differential equations. The first one is the homogeneous solution, and the other one is the particular solution depending on the given input. For the homogeneous solution of the difference equation, the right-hand side of the equation is set to zero, i.e.,

$$\sum_{k=0}^{N} a_k y[n-k] = 0 \tag{3.51}$$

from which we can write the characteristic equation

$$\sum_{k=0}^{N} a_k r^k = 0$$

which has N roots and the roots are indicated by r_i. Using these N roots, the solution of the equation can be written as

$$y_h[n] = \sum_{i=1}^{N} c_i r^i$$

where the coefficients c_i are calculated using the initial conditions.

Table 3.2 Solution table for difference equations

Input $x[n]$	Particular solution $y_p[n]$
K (constant number)	C (constant number)
n	$c_1 n + c_2$
α^n	$c\alpha^n$
n^k	$c_1 n^k + c_2 n^{k-1} + \ldots + c_k$
$\cos(\Omega n + \phi)$	$c_1 \cos(\Omega n) + c_2 \sin(\Omega n)$

Example 3.41 Find the homogeneous solution of the difference equation

$$y[n] - \frac{9}{16}y[n-2] = x[n-1]$$

Solution 3.41 The characteristic equation of

$$y[n] - \frac{9}{16}y[n-2] = 0$$

can be written as

$$r^n - \frac{9}{16}r^{n-2} = 0 \rightarrow r^2 - \frac{9}{16} = 0$$

whose roots are $r_{1,2} = \pm\frac{3}{4}$, and using the roots, we can write the homogeneous solution as

$$y_h[n] = c_1 \left(\frac{3}{4}\right)^n + c_2 \left(-\frac{3}{4}\right)^n$$

For the particular solutions of difference equations, we use a table as in the differential equations. Table 3.2 is used for the particular solutions of difference equations.

Example 3.42 Find the particular solution, $y_p[n]$, of

$$y[n] + \frac{1}{4}y[n-1] = x[n] + 2x[n-2]$$

for the input $x[n] = \left(\frac{1}{2}\right)^n$.

Solution 3.42 For system input α^n, the particular solution is of the form $c\alpha^n$, and if the particular solution is substituted in the equation, we get

$$c\alpha^n + \frac{c}{4}\alpha^{n-1} = \left(\frac{1}{2}\right)^n + 2\left(\frac{1}{2}\right)^{n-2} \rightarrow \left(c + \frac{c}{4\alpha}\right)\alpha^n = 9\left(\frac{1}{2}\right)^n$$

from which we find $\alpha = \frac{1}{2}$ and $c = 6$, and using the found numbers, the particular solution can be written as

$$y_p[n] = 6\left(\frac{1}{2}\right)^n$$

The total solution is obtained summing the homogeneous and particular solutions, i.e.,

$$y[n] = y_k[n] + y_p[n]$$

Example 3.43 Find the total solution of

$$y[n] - \frac{1}{4}y[n-1] = x[n] \tag{3.52}$$

for input $x[n] = \left(\frac{1}{2}\right)^n u[n]$ and initial condition $y[-1] = 8$.

Solution 3.43 First, let us find the homogeneous solution. For the homogeneous solution, the characteristic equation can be written as

$$r^n - \frac{1}{4}r^{n-1} = 0$$

from which we find

$$r = \frac{1}{4}$$

and homogeneous solution can be written as

$$y_h[n] = c\left(\frac{1}{4}\right)^n$$

The particular solution for the input

$$x[n] = \left(\frac{1}{2}\right)^n u[n]$$

is of the form

$$y_p[n] = k\left(\frac{1}{2}\right)^n u[n]$$

If we substitute the particular solution in (3.52), we get

$$k\left(\frac{1}{2}\right)^n u[n] - \frac{k}{4}\left(\frac{1}{2}\right)^{n-1} u[n-1] = \left(\frac{1}{2}\right)^n u[n]$$

from which for $n \geq 1$ we find $k = 2$, and for $n = 0$, we find $k = 1$. The particular solution can be written as

$$y_p[n] = \left(\frac{1}{2}\right)^n \delta[n] + 2\left(\frac{1}{2}\right)^n u[n-1]$$

The total solution is the sum of homogeneous and particular solutions, and it is obtained as

$$y[n] = c\left(\frac{1}{4}\right)^n + \left(\frac{1}{2}\right)^n \delta[n] + 2\left(\frac{1}{2}\right)^n u[n-1]$$

where the coefficient c can be determined using the initial condition $y[-1] = 8$ as $c = 1$. Hence, the total solution can be written as

$$y[n] = \left(\frac{1}{4}\right)^n + \left(\frac{1}{2}\right)^n \delta[n] + 2\left(\frac{1}{2}\right)^n u[n-1].$$

Exercise The input-output relationship of a digital system is indicated by the difference equation

$$y[n] = \rho y[n-1] + x[n]$$

Show that the system has linearity and time-invariance properties.

Problems

1. For input $x[n] = u[n+3]$, find the solution of

$$y[n] = \rho y[n-1] + x[n]$$

2. Find the impulse response of

$$y[n] + \frac{1}{2}y[n-1] - \frac{1}{3}y[n-3] = x[n] + 2x[n-2]$$

3. For an LTI system, for input $x(t) = e^{-\alpha t}u(t)$ and for the impulse response $h(t) = e^{-\beta t}u(t)$, find system output $y(t)$.
4. For an LTI system, for input $x(t) = \delta(t+1) - 2\delta(t-1) + \delta(t-3)$ and for the impulse response $h(t) = u(-t-2)$, find system output $y(t)$.

5. For an LTI system, for input $x(t) = e^{-2t}u(t)$ and for the impulse response $h(t) = u(-t - 2)$, find system output $y(t)$.

6. For a linear system, the input $\delta(t - \tau)$ produces the output

$$h_\tau(t) = u(t - \tau) - u(t - 2\tau)$$

(a) Determine whether the system is time-invariant or not.
(b) Determine whether the system is causal or not.
(c) Find the system outputs for the inputs

$$x(t) = u(t - 1) - u(t - 2) \quad x(t) = e^{-t}u(t)$$

7. Two LTI systems have the impulse responses $h_1[n] = \sin(4n)$ and $h_2[n] = a^n u[n]$. Find the outputs of the systems for the input

$$x[n] = \delta[n] - a\delta[n - 1], \quad |a| < 1$$

8. The impulse responses of some systems are given as

(a) $h[n] = \left(\frac{1}{3}\right)^n u[n]$
(b) $h[n] = (3)^n u[-n + 1]$
(c) $h[n] = \left(\frac{1}{3}\right)^n u[n]$
(d) $h(t) = e^{-3t}u(t - 2)$
(e) $h(t) = e^t u(-t - 2)$
(f) $h(t) = e^{-4|t|}$

Determine the causality and stability properties of these systems.

9. The input-output relationship of a system is given as

$$y[n] + 2y[n - 1] = x[n] + 2x[n - 3]$$

Find the impulse response of the system.

10. Find the homogeneous solutions of the equations:

(a) $\frac{d^2 y(t)}{dt} + \frac{3dy(t)}{dt} + 2y(t) = 0, \quad y(0) = 0, \quad y'(0) = 2$
(b) $y[n] + 2y[n - 1] + \frac{1}{8}y[n - 2] = 0, \quad y[0] = 1, \quad y[-1] = 6$

11. Find the convolutions of the functions:

(a) $f(t) = \delta(t - 1)$ and $g(t) = \delta(2t + 3)$
(b) $f(t) = \delta(t^2 - 1)$ and $g(t) = \delta\left(t - \frac{1}{2}\right)$
(c) $f(t) = u(t)$ and $g(t) = u(t - 2)$
(d) $f(t) = u(t)$ and $g(t) = r(t)$

12. Find the convolutions of the functions shown in Fig. P3.1.

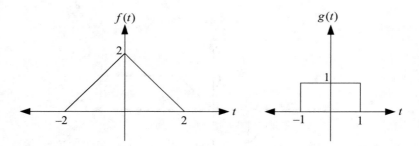

Fig. P3.1 The graphs of $f(t)$ and $g(t)$ for Problem 12

13. Find the convolutions of the functions shown in Fig. P3.2.

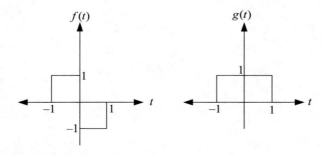

Fig. P3.2 The graphs of $f(t)$ and $g(t)$ for Problem 13

14. Find the energy and power of the signals:

 (a) $f(t) = e^{-\alpha t}$, $\alpha > 0$
 (b) $f(t) = e^{-\alpha t}u(t)$, $\alpha > 0$
 (c) $f(t) = u(t) - 2u(t - 10)$
 (d) $f(t) = r(t) - 2r(t - 4) + r(t - 10)$
 (e) $f(t) = \sin(2t)u(t)$
 (f) $f(t) = \sin(2u(t))$

15. Find the periods of the following signals, and calculate the powers of the signals in one period:

 (a) $f(t) = \sin(2\pi t)$
 (b) $f(t) = \sin(2\pi t + 1)$
 (c) $f(t) = \sin(2\pi t) + \cos(3\pi t)$
 (d) $f[n] = \cos(2\pi n)$
 (e) $f[n] = \cos(2\pi n) + \cos(3\pi n)$

16. Calculate the energy and power of the signal shown in Fig. P3.3.

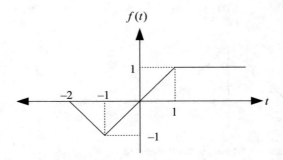

Fig. P3.3 The graph of $f(t)$ for Problem 16

Chapter 4
Fourier Analysis of Continuous-Time Signals

In this chapter, we explain the Fourier series representation of continuous-time signals. The history of the trigonometric sum goes back to the Babylonians. First, Euler examined the vibrating string and discovered that the frequencies produced by the string were harmonics of each other. Following Euler, the French scientist Fourier, born in 1768, worked on trigonometric series and showed that periodic signals could be written as the sum of harmonically related sinusoidal signals.

Fourier later showed that non-periodic signals can be written as the integral of sinusoidal signals that are not harmonic with each other. Fourier's work did not attract much attention during his lifetime. After his death, his work revolutionized the world of mathematics, and his studies are used in many areas of the engineering sciences. During Fourier's lifetime, many talented scientists like him inspected his work. Some of these scientists are Laplace, Lagrange, Euler, and Monge. Fourier's contributions to the science of mathematics led to important studies in many areas of engineering. Many new designs are achieved by examining the properties of signals in the frequency domain. In particular, music, medical engineering, and communication engineering were accelerated by Fourier's contributions.

In this chapter, first, we define the periodic signal and learn how to calculate the period of periodic signals. Next, we deal with the Fourier series representations of continuous-time signals. That is, we explain how to express periodic signals as the sum of complex sinusoidal signals. In the sequel, we focus on the Fourier integral representations of non-periodic signals. The Fourier integral representation of non-periodic signals is a mathematical representation method developed using periodic signals. Following the Fourier integral subject, we consider the Fourier transform, which is the complex representation of the Fourier integral. The Fourier transform expression can be considered to be the most important subject of signal processing, and many methods developed are based on the Fourier transform.

O. Gazi, *Principles of Signals and Systems*, https://doi.org/10.1007/978-3-031-17789-7_4

4.1   Fourier Series Representations of Continuous-Time Periodic Signals

Continuous-Time Periodic Signal
The signal $f(t)$ is a periodic signal if it satisfies

$$f(t) = f(t + kT), \quad k \in Z, \quad T \in R \tag{4.1}$$

where T is the fundamental period of $f(t)$. Any multiple of the fundamental period is another period of the function. The frequency of a signal whose principal period is T is calculated as

$$f = \frac{1}{T} \tag{4.2}$$

Angular frequency is defined as

$$w = 2\pi f \tag{4.3}$$

which has the unit of radian per second, i.e., $\frac{\text{radian}}{\text{s}}$.

Example 4.1 Find the period of $f(t) = \cos(200\pi t)$.

Solution 4.1 Using $f(t) = f(t + T)$ for the given function, we get

$$\cos(200\pi t) = \cos(200\pi(t + T))$$

which can be written as

$$\cos(200\pi t + 2k\pi) = \cos(200\pi t + 200\pi T), \quad k \in Z$$

where using $k = 1$ we calculate the fundamental period as

$$2\pi = 200\pi T \rightarrow T = \frac{1}{100}.$$

Property Let the period of $f(t)$ be T_1 and the period of $g(t)$ be T_2. The period of

$$h(t) = af(t) + bg(t)$$

can be obtained as

$$T = \text{smallest common multiples of } T_1 \text{ and } T_2.$$

Example 4.2 Find the period of $f(t) = \cos\left(\frac{\pi}{2}t\right) + \sin\left(\frac{\pi}{7}t\right)$.

Solution 4.2 The period of $\cos\left(\frac{\pi}{2}t\right)$ is

$$\frac{2\pi}{\frac{\pi}{2}} = 4$$

and the period of $\sin\left(\frac{\pi}{7}t\right)$ is

$$\frac{2\pi}{\frac{\pi}{7}} = 14.$$

The smallest common multiple of 14 and 4 is 28 which is the period of $f(t)$.

Property If the period of $f(t)$ is T, then the period of $f(kT)$, $k \in R$ is $\frac{T}{k}$.

Fourier Series Representations of Continuous-Time Periodic Signals
In this section, we will see how to write any periodic signal in terms of base signals. As it is known in linear algebra, any vector can be written as a combination of unit (base) vectors, i.e., $v = c_1 e_1 + c_2 e_2 + c_3 e_3 + \ldots$ where $e_1, e_2, e_3, \ldots$ are basis vectors, $c_1, c_2, c_3, \ldots$ are real numbers (coefficients). Let us give a physical example to warm up the subject further. Let us take a look at the cake making. The basic ingredients needed to make a cake can be summarized as flour, sugar, salt, oil, and eggs. Here, flour, sugar, salt, oil, and eggs are the basic ingredients used for cake making. The basic form of the cake is obtained by mixing these basic ingredients in certain proportions. Bread products can also be obtained by mixing these basic materials in certain proportions. Then, any product can be obtained using certain portions of basic materials.

A similar logic applies to periodic signals. Any periodic signal can be obtained by adding cosine, sine signals, and a real number in certain proportions. Cosine and sine signals can be thought of as the basic materials of periodic signals. Just as matter is formed by the combination of molecules in various proportions, periodic signals are obtained by multiplying the cosine and sine functions with certain coefficients and summing them. For example, a periodic square wave can be written as the sum of cosine and sine signals.

Theorem Let $f(t)$ be a periodic signal such that

$$f(t) = f(t + kT), \quad k \in Z, \quad T \in R \tag{4.4}$$

where T is the fundamental period of $f(t)$. Fourier series representation of $f(t)$ is

$$f(t) = A[0] + \sum_{k=1}^{\infty} A[k] \cos\left(k\frac{2\pi}{T}t\right) + \sum_{k=1}^{\infty} B[k] \sin\left(k\frac{2\pi}{T}t\right). \tag{4.5}$$

where $A[0]$, $A[k]$, and $B[k]$ are the Fourier series coefficients and they are calculated as

$$A[0] = \frac{1}{T} \int_T f(t)dt \quad A[k] = \frac{2}{T} \int_T f(t) \cos\left(k\frac{2\pi}{T}t\right) dt$$

$$B[k] = \frac{2}{T} \int_T f(t) \sin\left(k\frac{2\pi}{T}t\right) dt$$

(4.6)

Fourier series representation of $f(t)$ with some minor differences can also be given as

$$f(t) = \sqrt{\frac{1}{T}}A[0] + \sqrt{\frac{2}{T}} \sum_{k=1}^{\infty} A[k] \cos\left(k\frac{2\pi}{T}t\right) + \sqrt{\frac{2}{T}} \sum_{k=1}^{\infty} B[k] \sin\left(k\frac{2\pi}{T}t\right)$$

$$A[0] = \sqrt{\frac{1}{T}} \int_T f(t)dt$$

$$A[k] = \sqrt{\frac{2}{T}} \int_T f(t) \cos\left(k\frac{2\pi}{T}t\right) dt$$

$$B[k] = \sqrt{\frac{2}{T}} \int_T f(t) \sin\left(k\frac{2\pi}{T}t\right) dt$$

In this book, we will use (4.5) and (4.6) for Fourier series representation. The Fourier series representation takes a simpler form for odd and even signals.

Even Signal
If $f(t)$ satisfies $(t) = f(-t)$, then $f(t)$ is an even signal. The graphs of even signals are symmetric with respect to the vertical axis.

Example 4.3 The graph of $f(t)$ is depicted in Fig. 4.1 where it is seen that the signal is symmetric with respect to the vertical axis.

Fig. 4.1 Graph of $f(t)$ for Example 4.3

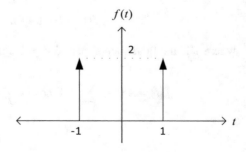

Fig. 4.2 Graph of $f(t)$ for
Example 4.4

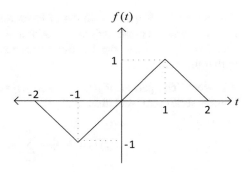

Fig. 4.3 Graph of $f(t)$ for
Example 4.5

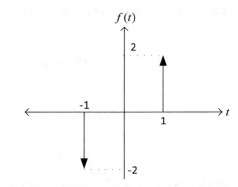

Odd Signal

If $f(t)$ satisfies $f(-t) = -f(t)$, then $f(t)$ is an odd signal. The graphs of odd signals are symmetric with respect to the origin.

Example 4.4 The graph of the odd signal $f(t)$ is depicted in Fig. 4.2 where it is seen that the graph is symmetric with respect to the origin.

Example 4.5 The graph of an odd signal is depicted in Fig. 4.3.

Any signal can be odd or even or have neither of these two properties. It is possible to write any arbitrary signal as the sum of an odd and an even signal, i.e.,

$$f(t) = g(t) + k(t)$$

where $f(t)$ is an arbitrary signal, $g(t)$ is an even signal, and $k(t)$ is an odd signal. The signals $g(t)$ and $k(t)$ can be obtained from $f(t)$ as

$$g(t) = \frac{f(t) + f(-t)}{2} \qquad k(t) = \frac{f(t) - f(-t)}{2} \qquad (4.7)$$

For even signal $f(t)$, we have $k(t) = 0$, and for odd signal $f(t)$, we have $g(t) = 0$.

Fourier Series Representations of Even and Odd Signals
Fourier series representations of odd and even signals are simpler than other
functions. Let us see the Fourier series representations of odd and even signals
with a theorem.

Theorem The period of $f(t)$ is T, and $f(t)$ is an even signal, i.e., $f(t) = f(-t)$. The
Fourier series representation of $f(t)$ is

$$f(t) = A[0] + \sum_{k=1}^{\infty} A[k] \cos\left(k\frac{2\pi}{T}t\right)$$

where the coefficients are calculated as

$$A[0] = \frac{2}{T}\int_0^{\frac{T}{2}} f(t)dt$$

$$A[k] = \frac{4}{T}\int_0^{\frac{T}{2}} f(t) \cos\left(k\frac{2\pi}{T}t\right)dt \tag{4.8}$$

$$B[k] = 0, \ k = 0,1,2,\ldots$$

where it is seen that $B[k]$ equals to zero, i.e., for an even signal, only $A[k]$ can have
non-zero values, and no need to calculate $B[k]$. When calculating $A[k]$ for even
signals, the integral is taken from 0 to $T/2$, and unlike the previous ones, there is a
coefficient $4/T$ at the beginning of the integral.

Let us now examine the case of odd signals. Assume that $f(t)$ is an odd periodic
signal, such that $f(-t) = -f(t)$. The Fourier series expansion of the signal is

$$f(t) = \sum_{k=1}^{\infty} B[k] \cos\left(k\frac{2\pi}{T}t\right)$$

where the coefficients are calculated as

$$A[0] = 0$$
$$A[k] = 0, k = 1,2,\ldots$$

$$B[k] = \frac{4}{T}\int_0^{\frac{T}{2}} f(t) \sin\left(k\frac{2\pi}{T}t\right)dt \tag{4.9}$$

For odd signals, the coefficients $A[0]$ and $A[k]$ equal to zero, and for the calculation of $B[k]$, the integral is evaluated from 0 to $T/2$, and unlike the previous ones,
coefficient $4/T$ is used in front of the integral.

Fig. 4.4 One period of a
periodic signal

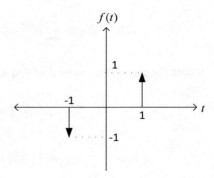

Example 4.6 One period of a periodic signal is given in Fig. 4.4. The period of the
signal is $T = 4$. Obtain the Fourier series representation of the signal.

Solution 4.6 The signal is symmetric about the origin, and it is an odd signal. There
is no need to calculate the coefficients $A[0]$ and $A[k]$; since the coefficients $A[0]$ and
$A[k]$ equal to zero, it is sufficient to calculate the coefficient $B[k]$.

Considering Fig. 4.4, we can obtain the mathematical expression of the signal for
$0 \leq t \leq 2$ as

$$f(t) = -\delta(t+1) + \delta(t-1) \rightarrow 0 \leq t < 2 \quad f(t) = \delta(t-1)$$

The coefficients $B[k]$ can be calculated as

$$B[k] = \frac{4}{T} \int_{0}^{\frac{T}{2}} f(t) \sin\left(k\frac{2\pi}{T}t\right) dt \rightarrow B[k] = \frac{4}{4} \int_{0}^{2} \delta(t-1) \sin\left(k\frac{2\pi}{T}t\right) dt$$

$$\rightarrow B[k] = \sin\left(k\frac{\pi}{2}\right)$$

The Fourier series representation of $f(t)$ using $B[k]$ can be obtained as

$$f(t) = \sum_{k=1}^{\infty} B[k] \cos\left(k\frac{2\pi}{T}t\right)$$

$$= \sum_{k=1}^{\infty} \sin\left[k\frac{\pi}{2}\right] \cos\left(k\frac{\pi}{2}t\right)$$

Complex Fourier Series Representation
Let $f(t)$ be a periodic signal such that $f(t) = f(t+T)$; Fourier series representation of
$f(t)$ can be written as

$$f(t) = A[0] + \sum_{k=1}^{\infty} A[k] \cos \left(k\frac{2\pi}{T}t\right) + \sum_{k=1}^{\infty} B[k] \sin \left(k\frac{2\pi}{T}t\right)$$

where the coefficients are calculated as

$$A[0] = \frac{1}{T} \int_T f(t)dt$$

$$A[k] = \frac{2}{T} \int_T f(t) \cos \left(k\frac{2\pi}{T}t\right)dt \quad B[k] = \frac{2}{T} \int_T f(t) \sin \left(k\frac{2\pi}{T}t\right)dt$$

where the coefficients $A[0]$, $A[k]$, and $B[k]$ are real numbers. It is possible to combine these three coefficients under a single coefficient, and it is possible to express the three formulas summed in the Fourier series expansion of $f(t)$ with a single formula. The simpler expression is called the complex Fourier series representation of the periodic function and is given as

$$f(t) = \sum_{k=-\infty}^{\infty} F[k]e^{jk\frac{2\pi}{T}t}$$

$$F[k] = \frac{1}{T} \int_T f(t)e^{-jk\frac{2\pi}{T}t}dt \tag{4.10}$$

where $F[k]$ are called the complex Fourier series coefficients of the function $f(t)$. Fourier series representation of periodic signals with complex coefficients can alternatively be defined as

$$f(t) = \frac{1}{\sqrt{T}} \sum_{k=-\infty}^{\infty} F[k]e^{jkw_0t}$$

$$F[k] = \frac{1}{\sqrt{T}} \int_T f(t)e^{-jkw_0t}dt \tag{4.11}$$

where $F[k]$ are complex numbers.

The formulas in (4.10) can be written in a more simplified form as

$$w_0 = \frac{2\pi}{T}$$

$$f(t) = \sum_{k=-\infty}^{\infty} F[k]e^{jkw_0t} \qquad F[k] = \frac{1}{T}\int_T f(t)e^{-jkw_0t}dt$$

The relationships between $F[k]$ and $A[k]$, $B[k]$ are given as

$$F[0] = A[0] \qquad F[k] = \frac{1}{2}(A[k] - jB[k]) \quad k \neq 0 \tag{4.12}$$

Proof   According to our claim, we can write

$$A[0] + \sum_{k=1}^{\infty} A[k]\cos\left(k\frac{2\pi}{T}t\right) + \sum_{k=1}^{\infty} B[k]\sin\left(k\frac{2\pi}{T}t\right) = \sum_{k=-\infty}^{\infty} F[k]e^{jk\frac{2\pi}{T}t} \tag{4.13}$$

Starting from the left-hand side of (4.13), let us try to obtain the right-hand side of (4.13). For the simplicity of the proof, let us take the period of the function as $T = 2\pi$.

For the rest of the proof, we will use the formulas

$$e^{jkt} = \cos(kt) + j\sin(kt) \qquad e^{-jkt} = \cos(kt) - j\sin(kt)$$

$$\cos(kt) = \frac{1}{2}\left(e^{jkt} + e^{-jkt}\right) \qquad \sin(kt) = \frac{1}{2j}\left(e^{jkt} - e^{-jkt}\right) \tag{4.14}$$

$$f(t) = A[0] + \sum_{k=1}^{\infty}[A[k]\cos(kt) + B[k]\sin(kt)] \quad T = 2\pi \tag{4.15}$$

If the mathematical expressions given in (4.14) for $\cos(kt)$ and $\sin(kt)$ are substituted in (4.15), we obtain

$$\begin{aligned} f(t) &= A[0] + \sum_{k=1}^{\infty}\left(\frac{1}{2}(A[k] - jA[k])e^{jkt} + \frac{1}{2}(A[k] + jB[k])e^{-jkt}\right) \\ &= F[0] + \sum_{k=1}^{\infty}\left(F[k]e^{jkt} + K[k]e^{-jkt}\right) \end{aligned} \tag{4.16}$$

where

$$F[k] = \frac{1}{2}(A[k] - jB[k]), \quad F[0] = A[0], \quad K[k] = \frac{1}{2}(A[k] + jB[k]) \tag{4.17}$$

If $A[k]$ and $B[k]$ in (4.6) are substituted into (4.17), we get

$$F[k] = \frac{1}{2\pi} \int_{-\pi}^{\pi} f(t)[\cos(kt) - j\sin(kt)]dt$$

$$= \frac{1}{2\pi} \int_{-\pi}^{\pi} f(t)e^{-jkt}dt \qquad (4.18)$$

$$K[k] = \frac{1}{2\pi} \int_{-\pi}^{\pi} f(t)[\cos(kt) + j\sin(kt)]dt$$

$$= \frac{1}{2\pi} \int_{-\pi}^{\pi} f(t)e^{jkt}dt \qquad (4.19)$$

In (4.18) and (4.19), we see that $F[k] = K[-k]$, and using $F[k] = K[-k]$ in (4.16), we obtain

$$f(t) = F[0] + \sum_{k=1}^{\infty} \left(F[k]e^{jkt} + F[-k]e^{-jkt} \right)$$

$$= \sum_{k=-\infty}^{\infty} F[k]e^{jkt} \qquad (4.20)$$

where

$$F[k] = \frac{1}{2\pi} \int_{-\pi}^{\pi} f(t)e^{-jkt}dt, k \in Z \qquad (4.21)$$

The proof is complete.

Note In general, the Fourier series representation of the periodic signal $f(t)$ with period T is

$$f(t) = c_1 \sum_{k=-\infty}^{\infty} F[k]e^{jk\frac{2\pi}{T}t} \qquad (4.22)$$

where $F[k]$ is calculated as

$$F[k] = c_2 \int_{T} f(t)e^{-jk\frac{2\pi}{T}t}dt \qquad (4.23)$$

and coefficients c_1 and c_2 satisfy

$$c_1 \times c_2 = \frac{1}{T} \qquad (4.24)$$

That is, if $c_1 = 1$, then c_2 must be chosen as $c_2 = \frac{1}{T}$.
If $c_1 = \frac{1}{\sqrt{T}}$, then c_2 equals to $c_2 = \frac{1}{\sqrt{T}}$.
If $c_1 = \frac{1}{T}$, then c_2 must be chosen as $c_2 = 1$.

Before considering the examples related to the subject, let us give a property about the Fourier series coefficients. This property makes it easy to solve some problems. Therefore, we find it appropriate to give it immediately.

Property Let $f(t)$ and $g(t)$ be two periodic signals with a common period T, and the relationship between these two signals is given as

$$g(t) = \frac{df(t)}{dt} \qquad (4.25)$$

The relationship between the complex Fourier series coefficients $G[k]$ of the function $g(t)$ and the complex Fourier coefficients $F[k]$ of the function $f(t)$ is as

$$G[k] = jkw_0 F[k] \quad w_0 = \frac{2\pi}{T} \qquad (4.26)$$

Proof The Fourier series representation of $f(t)$ can be written as

$$w_0 = \frac{2\pi}{T} \quad f(t) = \sum_{k=-\infty}^{\infty} F[k] e^{jkw_0 t}$$

Taking the derivative of $f(t)$ in (4.25), we obtain

$$g(t) = \frac{df(t)}{dt} \rightarrow g(t) = \sum_{k=-\infty}^{\infty} jkw_0 F[k] e^{jkw_0 t} \qquad (4.27)$$

Fourier series representation of $g(t)$ can be written as

$$g(t) = \sum_{k=-\infty}^{\infty} G[k] e^{jkw_0 t} \qquad (4.28)$$

Equating the right-hand sides of (4.27) and (4.28), we get

$$G[k] = jkw_0 F[k]$$

Fig. 4.5 Graph of *f(t)* for
Example 4.7

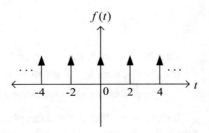

Example 4.7 First, write the periodic signal in Fig. 4.5 mathematically, and then obtain the Fourier series representation of the signal.

Solution 4.7 It is seen from Fig. 4.5 that the period of the signal is 2. Periodic signals are obtained by shifting one period of the signal to the right and left by multiples of the period and summing them. Accordingly, the mathematical expression of the signal is

$$f(t) = \sum_{k=-\infty}^{\infty} \delta(t - 2k).$$

To find the Fourier series coefficients of the signal, let us consider its one period. The Fourier series coefficients can be calculated using

$$F[k] = \frac{1}{T} \int_{-\frac{T}{2}}^{\frac{T}{2}} f(t)e^{-jkw_0t}dt \quad w_0 = \frac{2\pi}{T}.$$

on the interval $-1 \le t \le 1$ for $T = 2$ as

$$F[k] = \frac{1}{2} \int_{-1}^{1} \delta(t)e^{-jkw_0t}dt \quad w_0 = \pi$$

$$= \frac{1}{2}e^0$$

$$= \frac{1}{2}.$$

Using the Fourier series coefficients, the Fourier series representation of the signal can be obtained as

$$f(t) = \sum_{k=-\infty}^{\infty} F[k]e^{jkw_0t} \rightarrow f(t) = \frac{1}{2} \sum_{k=-\infty}^{\infty} e^{jk\pi t}.$$

If we equate the Fourier series representation of the signal $f(t)$ with its mathematical expression, we get the interesting equation

$$\sum_{k=-\infty}^{\infty} \delta(t-2k) = \frac{1}{2} \sum_{k=-\infty}^{\infty} e^{jk\pi t}. \tag{4.29}$$

The equality in (4.29) is obtained for the period value $T = 2$, and the equality (4.29) can be written for any value of T as

$$\sum_{k=-\infty}^{\infty} \delta(t-2k) = \frac{1}{T} \sum_{k=-\infty}^{\infty} e^{jk\pi t} \qquad w_0 = \frac{2\pi}{T}.$$

Example 4.8 Find the real Fourier series coefficients of the signal given in Fig. 4.5, and obtain the Fourier series representation of the signal using real Fourier series coefficients.

Solution 4.8 The real Fourier series coefficients can be calculated as

$$A[0] = \frac{1}{T} \int_T f(t)dt \rightarrow A[0] = \frac{1}{2} \int_{-1}^{1} \delta(t)dt \rightarrow A[0] = \frac{1}{2}$$

$$A[k] = \frac{2}{T} \int_T f(t) \cos\left(k\frac{2\pi}{T}t\right)dt \rightarrow A[k] = \frac{2}{2} \int_{-1}^{1} \delta(t) \cos\left(k\frac{2\pi}{T}t\right)dt \rightarrow A[k] = 1$$

$$B[k] = \frac{2}{T} \int_T f(t) \sin\left(k\frac{2\pi}{T}t\right)dt \rightarrow B[k] = \frac{2}{2} \int_{-1}^{1} \delta(t) \sin\left(k\frac{2\pi}{T}t\right)dt \rightarrow B[k] = 0$$

$$f(t) = A[0] + \sum_{k=1}^{\infty} A[k] \cos\left(k\frac{2\pi}{T}t\right) \rightarrow f(t) = \frac{1}{2} + \sum_{k=1}^{\infty} \cos(k\pi t).$$

Example 4.9 Obtain the Fourier series representation of the signal shown in Fig. 4.6. Obtain real and complex Fourier series coefficients separately.

Solution 4.9 It is seen from Fig. 4.6 that the signal is a periodic square wave with a period of 2π. First, let us calculate the real Fourier series coefficients. The coefficients $A[k]$ are calculated as

Fig. 4.6 Graph of $f(t)$ for
Example 4.9

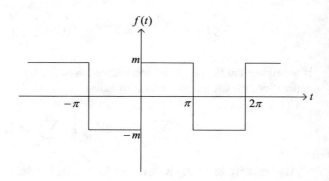

$$A[0] = \frac{1}{2\pi} \int\limits_{-\pi}^{\pi} f(t)dt \qquad\qquad A[k] = \frac{1}{\pi} \int\limits_{-\pi}^{\pi} f(t)\cos(kt)dt$$

$$= \frac{1}{2\pi} \left[\int\limits_{-\pi}^{0} f(t)dt + \int\limits_{0}^{\pi} f(t)dt \right] \qquad = \frac{1}{\pi} \left[\int\limits_{-\pi}^{0} (-m)\cos(kt)dt \right.$$

$$= \frac{1}{2\pi}[(-m)\pi + m\pi] \qquad\qquad \left. + \int\limits_{0}^{\pi} m\cos(kt)dt \right]$$

$$= 0 \qquad\qquad = \frac{1}{\pi}\left[\left(-m\frac{\sin(kt)}{k}\Big|_{-\pi}^{0} + (m\frac{\sin(kt)}{k}\Big|_{0}^{\pi}\right)\right]$$

$$= \frac{1}{2\pi}[(-m)\pi + m\pi] = 0$$

where it is seen that $A[0]$ and $A[k]$ equal to zero. Since the signal is an odd signal, without doing any calculations, we could directly write that $A[k]$ coefficients are equal to zero. The coefficients $B[k]$ are calculated as

$$B[k] = \frac{1}{\pi} \int\limits_{-\pi}^{\pi} f(t)\sin(kt)dt$$

$$= \frac{1}{\pi} \left[\int\limits_{-\pi}^{0} (-m)\sin(kt)dt + \int\limits_{0}^{\pi} m\sin(kt)dt \right]$$

$$= \frac{1}{\pi} \left[m\frac{\cos(kt)}{k}\Big|_{-\pi}^{0} + (-m)\frac{\cos(kt)}{k}\Big|_{0}^{\pi} \right]$$

$$= \frac{m}{k\pi}[\cos(0) - \cos(-k\pi) - \cos(k\pi) + \cos(0)]$$

$$= \frac{2m}{k\pi}(1 - \cos(k\pi)).$$

Using the calculated coefficients, we can write the Fourier series representation of the signal as

$$f(t) = A[0] + \sum_{k=1}^{\infty} A[k] \cos{(kt)} + B[k] \sin{(kt)}$$
$$= \sum_{k=1}^{\infty} \frac{2m}{k\pi} (1 - \cos{(k\pi)}) \sin{(kt)}.$$

Now let us calculate the complex Fourier series coefficients of the signal. For this purpose, we can use the formula used to calculate the complex coefficients, or we can directly use the formula showing the relationship between the complex coefficients and the real coefficients. For convenience, let us follow the second approach as

$$F[0] = A[0] \quad F[k] = \frac{1}{2}(A[k] - jB[k])$$

resulting in $F[0] = 0$ and $F[k] = -0.5jB[k]$. The complex coefficient Fourier series representation of the signal can be written as

$$f(t) = \sum_{k=-\infty}^{\infty} \frac{jm}{k\pi} (\cos{(k\pi)} - 1)e^{jkt}$$

To solve this problem, we directly used the formulas for the calculation of Fourier series coefficients. However, if we take the derivative of the signal first and calculate the Fourier coefficients for the derivative signal and then obtain the Fourier series coefficients of the $f(t)$ signal, we would deal with mathematically easier integrals. Since the derivative signal contains impulses, it is easier to integrate it.

In this problem, the integral calculation for $f(t)$ is not very difficult, but we recommend simplifying the integration operations by taking the derivatives of the signal so that the integral calculations can be made more easily. We will consider the solution with this method in the next example.

If we look at the Fourier series representation of the $f(t)$ signal, it is seen that the summation process goes on till infinity. In practice, the upper limit of the summation operator is a finite number. The limited summation results of the Fourier series expansion of $f(t)$ for $m = 2$ (amplitude) are displayed in Figs. 4.7, 4.8, and 4.9. It is seen from Figs. 4.7, 4.8, and 4.9 that the limited summation result looks more like a square wave as the number of summed terms increases.

The Fourier series representation for limited summation can be written as

$$f_N(t) = \sum_{k=1}^{N} \frac{2m}{k\pi} (1 - \cos{(k\pi)}) \sin{(kt)}$$

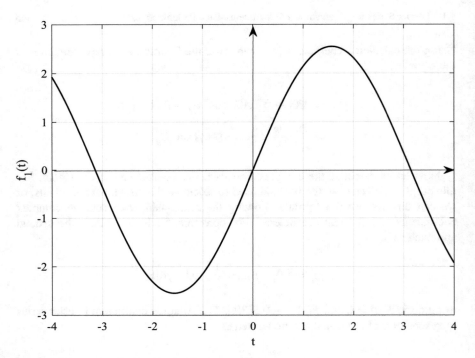

Fig. 4.7 The graph of $f_1(t) = \frac{4m}{\pi} \sin(t)$

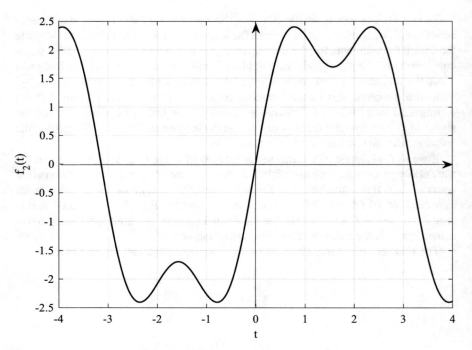

Fig. 4.8 The graph of $f_2(t) = \frac{4m}{\pi} \left(\sin(t) + \frac{1}{3} \sin(3t) \right)$

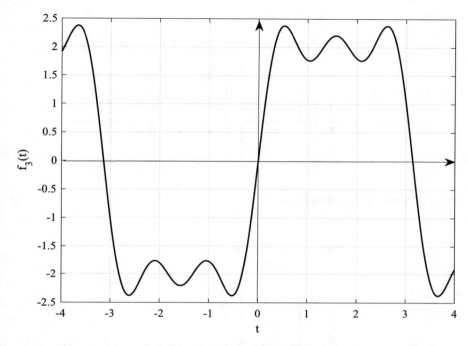

Fig. 4.9 The graph of $f_3(t) = \frac{4m}{\pi}\left(\sin(t) + \frac{1}{3}\sin(3t) + \frac{1}{5}\sin(5t)\right)$

which are plotted, for different N values and for the value of $m = 2$ (amplitude), in Figs. 4.7, 4.8, and 4.9.

It is seen from Figs. 4.7, 4.8, and 4.9 that the more N increases, the more the resulting graph gets closer to the square wave.

Example 4.10 Obtain the Fourier series representation of the signal shown in Fig. 4.10.

Solution 4.10 The Fourier series coefficients of $f(t)$ can be found directly, or by taking the derivative of the signal, the Fourier series coefficients of the derivative signal can be calculated first, and then the Fourier series coefficients of the original signal can be calculated using the Fourier series coefficients of the derivative signal.

Let $g(t)$ be the derivative of $f(t)$. The relationship between the complex Fourier series coefficients of the signal $f(t)$ and the complex Fourier series coefficients of the signal is

$$G[k] = jkw_0F[k], \quad w_0 = \frac{2\pi}{T} \tag{4.30}$$

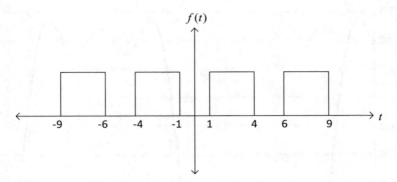

Fig. 4.10 Graph of $f(t)$ for Example 4.10

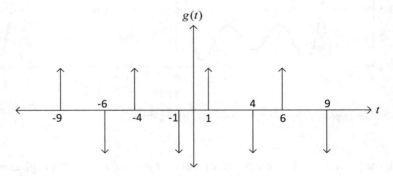

Fig. 4.11 The graph of the derivative function for Example 4.10

The graph of $g(t) = \frac{df(t)}{dt}$ is depicted in Fig. 4.11 where impulses have amplitude values 1 and -1.

The complex Fourier series coefficients of $g(t)$ can be calculated as

$$
\begin{aligned}
G[k] &= \frac{1}{T} \int_{T} g(t) e^{-jk\frac{2\pi}{T}t} dt \\
&= \frac{1}{5} \int_{-2}^{3} (-\delta(t+1) + \delta(t-1)) e^{-jk\frac{2\pi}{5}t} dt \\
&= \frac{1}{5} \left(-e^{jk\frac{2\pi}{5}} + e^{-jk\frac{2\pi}{5}} \right) \\
&= -\frac{2j}{5} \sin\left(\frac{2\pi k}{5}\right)
\end{aligned}
$$

which can be used to calculate the complex Fourier series coefficients of $f(t)$ as

$$F[k] = \frac{5}{k2\pi j}G[k] \rightarrow F[k] = -\frac{1}{k\pi}\sin\left(\frac{2\pi k}{5}\right)$$

It is seen from Fig. 4.11 that $g(t)$ is an odd signal. It is symmetric with respect to the origin. When finding the real Fourier series coefficients of $g(t)$, the odd signal property can also be used.

Example 4.11 Find the Fourier series coefficients of

$$f(t) = \sin\left(\frac{2\pi}{T}t\right)$$

Solution 4.11 The period of the sine signal given is T. Let us first write $f(t)$ as the sum of the exponential signals and then find the Fourier series coefficients by comparing it with the Fourier series representation formula. The sine signal

$$f(t) = \sin\left(\frac{2\pi}{T}t\right)$$

can be written in terms of the exponential signals as

$$f(t) = \frac{1}{2j}\left(e^{j\frac{2\pi}{T}t} - e^{-j\frac{2\pi}{T}t}\right) \tag{4.31}$$

The Fourier series representation of the signal $f(t)$ with period T is

$$\begin{aligned} f(t) &= \sum_{k=-\infty}^{\infty} F[k]e^{jk\frac{2\pi}{T}t} \\ &= \ldots + F[-1]e^{-j\frac{2\pi}{T}t} + F[0] + F[1]e^{j\frac{2\pi}{T}t} + \ldots \end{aligned} \tag{4.32}$$

By comparing (4.31) to (4.32), we obtain

$$F[k] = \begin{cases} \pm\dfrac{1}{2j} & k = \pm 1 \\ 0 & \text{otherwise} \end{cases}$$

Example 4.12 Find the Fourier series coefficients of

$$f(t) = \cos\left(\frac{2\pi}{T}t\right)$$

Solution 4.12 Following the same approach in the solution of the previous example, the Fourier series coefficients are found as

$$F[k] = \begin{cases} \dfrac{1}{2} & k = \pm 1 \\ 0 & \text{otherwise} \end{cases}$$

Convolution of Continuous-Time Periodic Signals

In this section, we will consider the convolution of the signals $x(t)$ and $y(t)$ with periods T. This convolution is called periodic convolution, and it is defined as

$$\begin{aligned} z(t) &= x(t) * y(t) \\ &= \int_0^T x(\tau)y(t-\tau)d\tau \end{aligned} \tag{4.33}$$

where the period of $z(t)$ is also T.

Let $X[k]$ and $Y[k]$ be the complex Fourier series coefficients of periodic signals $x(t)$ and $y(t)$. Accordingly, the relationship between the Fourier series coefficient $Z[k]$ of $z(t)$ and $X[k]$ and $Y[k]$ is as

$$Z[k] = TX[k]Y[k] \tag{4.34}$$

If the Fourier series coefficients of the $x(t)$ and $y(t)$ are known, first $Z[k]$ is calculated, and then $z(t)$ is obtained using

$$z(t) = \sum_k Z[k]e^{jk\frac{2\pi}{T}t} \tag{4.35}$$

Example 4.13 Given

$$x(t) = \cos\left(\frac{\pi}{2}t\right) \text{ and } y(t) = \sum_k p(t - 4k)$$

$$p(t) = \begin{cases} 1 & -1 \le t \le 1 \\ 0 & \text{otherwise.} \end{cases}$$

Find the periodic convolution of $x(t)$ and $y(t)$.

Solution 4.13 The common period of $x(t)$ and $y(t)$ is 4. The Fourier series coefficients of $x(t)$ can be found as

$$X[k] = \begin{cases} \dfrac{1}{2} & k = \pm 1 \\ 0 & \text{otherwise} \end{cases}$$

The Fourier series coefficients $y(t)$ can be calculated as

$$Y[k] = \frac{\sqrt{2\pi}}{T} \, \widehat{p}(w)\big|_{w=k\frac{2\pi}{4}}$$

where

$$\widehat{p}(w) = \frac{1}{\sqrt{2\pi}} \int\limits_{t=-\infty}^{\infty} p(t)e^{-jwt}dt$$

is the Fourier transform of $p(t)$ and it can be calculated as

$$\widehat{p}(w) = \frac{1}{\sqrt{2\pi}} \frac{\sin(w)}{w}$$

Using $\widehat{p}(w)$ in $Y[k]$, we obtain

$$Y[k] = \frac{1}{4} \frac{\sin\left(k\frac{\pi}{2}\right)}{k\frac{\pi}{2}}$$

from which $Z[k]$ can be calculated using (4.34) as

$$Z[k] = \begin{cases} \dfrac{1}{\pi} & k = \pm 1 \\ 0 & \text{otherwise} \end{cases}$$

from which $z(t)$ can be found as

$$z(t) = \frac{2}{\pi} \cos\left(\frac{\pi}{2}t\right)$$

Convergence of Fourier Series

Let $f(t)$ be a periodic signal such that $f(t) = f(t + T)$ where T is the fundamental period of the signal. A periodic signal must satisfy the following properties in order to have the Fourier series representation:

1. $f(t)$ is a piecewise continuous signal.
2. $f(t)$ has left and right derivatives at every point of the range in which it is defined.

It is possible to get the Fourier series representation of the periodic signal $f(t)$ that has both of the properties. If the signal does not satisfy one of the properties, it is not possible to get the Fourier series representation of the signal. The value of the Fourier series at the discontinuity points of $f(t)$ is equal to the mean value of the left and right limits of the signal at the discontinuity point.

Fig. 4.12 One period of a piecewise continuous periodic function

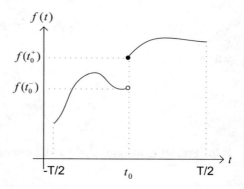

Example 4.14 $f(t)$ is a piecewise continuous periodic signal. One period of $f(t)$ is depicted in Fig. 4.12 where it is seen that the signal is discontinuous at a single point, and the signal has left and right derivative values at every point of the range in which it is defined.

Then, the signal has Fourier series representation, and it is given as

$$f(t) = A[0] + \sum_{k=1}^{\infty} A[k] \cos\left(k\frac{2\pi}{T}t\right) + \sum_{k=1}^{\infty} B[k] \sin\left(k\frac{2\pi}{T}t\right) \tag{4.36}$$

The value of the Fourier series representation of $f(t)$ at the discontinuity point is equal to the average of the left and right limits of the function at the discontinuity point, that is:

$$A[0] + \sum_{k=1}^{\infty} A[k] \cos\left(k\frac{2\pi}{T}t\right) + \sum_{k=1}^{\infty} B[k] \sin\left(k\frac{2\pi}{T}t\right)\bigg|_{t=t_0} = \frac{f\left(t_0^-\right) + f\left(t_0^+\right)}{2} \tag{4.37}$$

Parseval's Identity

The energy of the periodic signal $f(t)$ in its one period can be calculated as

$$\int_T |f(t)|^2 dt = T \sum_k |F[k]|^2 \tag{4.38}$$

where T is the period of the signal and $F[k]$ are the complex Fourier series coefficients whose squared absolute value can be written in terms of real Fourier series coefficients as

$$|F[k]|^2 = \frac{1}{2}\left(A^2[k] + B^2[k]\right) \tag{4.39}$$

We can prove Parseval's identity as

$$\int_T f^2(t)dt = \int_T \left(\sum_k F[k]e^{jk\frac{2\pi}{T}t}\right)^2 dt$$

$$= \int_T \left(\sum_k F[k]e^{jk\frac{2\pi}{T}t}\right)\left(\sum_l F[l]e^{jl\frac{2\pi}{T}t}\right)^* dt$$

$$= \int_T \left(\sum_k F[k]e^{jk\frac{2\pi}{T}t}\sum_l F[l]^* e^{-jl\frac{2\pi}{T}t}\right) dt$$

$$= \int_T \left(\sum_{k,l} F[k]F[l]^* e^{-j(k-l)\frac{2\pi}{T}t}\right) dt$$

$$= \int_T \left(\sum_k F[k]F[k]^* + \sum_{\substack{k,l \\ k \neq l}} F[k]F[l]^* e^{-j(k-l)\frac{2\pi}{T}t}\right) dt$$

$$= \sum_k \int_T F[k]F[k]^* dt + \sum_{\substack{k,l \\ k \neq l}} \int_T F[k]F[l]^* e^{-j(k-l)\frac{2\pi}{T}t} dt$$

where the expression

$$\sum_{\substack{k,l \\ k \neq l}} \int_T F[k]F[l]^* e^{-j(k-l)\frac{2\pi}{T}t} dt = \sum_{\substack{k,l \\ k \neq l}} \int_T F[k]F[l]^* \left(\cos\left((k-l)\frac{2\pi}{T}t\right)\right.$$

$$\left. -j\sin\left((k-l)\frac{2\pi}{T}t\right)\right) dt$$

equals to 0, since the integrals of the sine and cosine functions in one period equal zero. Thus, we obtain

$$\int_T f^2(t)dt = \sum_k \int_T F[k]F^*[k] dt$$

$$= T\sum_k F^2[k]$$

Example 4.15 Verify the Parseval's identity for the signal

$$f(t) = \sin\left(\frac{2\pi}{T}t\right)$$

Solution 4.15 The period of $f(t)$ is T. The energy of the signal in its one period can be calculated as

$$\int_T f^2(t)dt = \int_T \sin^2\left(\frac{2\pi}{T}t\right)dt$$

$$= \int_T \left(\frac{1 - \cos\left(\frac{4\pi}{T}t\right)}{2}\right)dt$$

$$= \int_T \frac{1}{2}dt - \frac{1}{2}\underbrace{\int_T \cos\left(\frac{4\pi}{T}t\right)dt}_{=0}$$

$$= \frac{T}{2}.$$

$f(t)$ can be written in terms of the exponential signals as in

$$f(t) = \sin\left(\frac{2\pi}{T}t\right)$$

$$= \frac{e^{j\frac{2\pi}{T}t} - e^{-j\frac{2\pi}{T}t}}{2j}$$

which can be compared to the Fourier series representation

$$f(t) = \sum_{k=-\infty}^{\infty} F[k]e^{jk\frac{2\pi}{T}t}$$

leading to

$$F[-1] = -\frac{1}{2j} \quad F[1] = \frac{1}{2j} \quad F[k] = 0, \quad k \neq -1 \quad k \neq 1$$

Accordingly,

$$\sum_k F^2[k] = \left(-\frac{1}{2j}\right)^2 + \left(\frac{1}{2j}\right)^2$$

$$= \frac{1}{2}.$$

Table 4.1 Properties of Complex Fourier Series Coefficients

$f(t) \overset{\text{CFSC}}{\leftrightarrow} F[k]$	$g(t) \overset{\text{CFSC}}{\leftrightarrow} G[k]$
$cf(t) + dg(t) \overset{\text{CFSC}}{\leftrightarrow} cF[k] + dG[k]$	$f(t - t_0) \overset{\text{CFSC}}{\leftrightarrow} e^{-jk\frac{2\pi}{T}t_0} F[k]$
$e^{jk_0\frac{2\pi}{T}t} f(t) \overset{\text{CFSC}}{\leftrightarrow} F[k - k_0]$	$f(\alpha t) \overset{\text{CFSC}}{\leftrightarrow} F[k], \ \alpha > 0$
$f(t) * g(t) \overset{\text{CFSC}}{\leftrightarrow} T F[k] G[k]$	$f(t)g(t) \overset{\text{CFSC}}{\leftrightarrow} F[k] * G[k]$
$\frac{f(t)+f(-t)}{2} \overset{\text{CFSC}}{\leftrightarrow} \text{Re}\,(F(k))$	$\frac{f(t)-f(-t)}{2} \overset{\text{CFSC}}{\leftrightarrow} j\text{Im}(F(k))$
$f(-t) \overset{\text{CFSC}}{\leftrightarrow} F[-k]$	$\int\limits_{\tau=-\infty}^{t} f(\tau)d\tau \overset{\text{CFSC}}{\leftrightarrow} \frac{F[k]}{jk\frac{2\pi}{T}}$ \quad if $F[0] = 0$

Table 4.2 Properties of Real Fourier Series Coefficients

$F[k] = F^*[-k]$	$\text{Re}(F[k]) = \text{Re}\,(F[-k])$
$\text{Im}(F[k]) = -\,\text{Im}\,(F[-k])$	$|F[k]| = |F[-k]|$
$\angle F[k] = -\,\angle F[-k]$	$f^*(t) \overset{\text{CFSC}}{\leftrightarrow} F[k]$

Thus, considering the obtained results, we can write

$$\int_T f^2(t)dt = T\sum_k F^2[k] \tag{4.40}$$

Properties of Fourier Series Coefficients

Let $f(t)$ and $g(t)$ be the two periodic signals with the same period T. Let the complex Fourier series coefficients of these signals be $F[k]$ and $G[k]$, respectively.

Properties involving the Fourier series coefficients of these signals are provided in Table 4.1 where CFSC means complex Fourier series coefficients, $\text{Re}(\cdot)$ is used to get the real part, and $\text{Im}(\cdot)$ is used to get the imaginary part.

For real-valued signal $f(t)$, the properties for Fourier series coefficients are given in Table 4.2.

4.2 Fourier Integral and Fourier Transform

In this section, we will review some of the basic topics of calculus. Let $f(t)$ be a signal that takes real values, and let t_0 be a real number.

The left-hand limit of $f(t)$ at t_0 is defined for $h \in R^+$ as

$$f(t_0^-) = \lim_{h \to 0} f(t_0 - h) \tag{4.41}$$

The right-hand limit of $f(t)$ at t_0 is defined for $h \in R^+$ as

$$f(t_0^+) = \lim_{h \to 0} f(t_0 + h) \tag{4.42}$$

If $f(t_0^-) = f(t_0^+)$, then $f(t)$ is said to be continuous at point t_0, and the value at point t_0 satisfies

$$f(t_0) = f(t_0^-) = f(t_0^+) \tag{4.43}$$

The left and right derivatives of $f(t)$ at t_0 are evaluated as

$$f'(t_0^-) = \lim_{h \to 0} \frac{f(t_0 - h) - f(t_0)}{h} \tag{4.44}$$

$$f'(t_0^+) = \lim_{h \to 0} \frac{f(t_0 + h) - f(t_0)}{-h} \tag{4.45}$$

If the left and right derivatives of $f(t)$ at t_0 are equal to each other, that is, if $f'(t_0^-) = f'(t_0^+)$, then $f(t)$ has a derivative at t_0, and the value of its derivative at t_0 is represented as $f'(t_0)$.

Example 4.16 The graph of a signal $f(t)$ is depicted in Fig. 4.13 where it is seen that the left and right limits of $f(t)$ are equal at t_0; this means that the signal is continuous at t_0.

Example 4.17 The graph of a signal $g(t)$ is depicted in Fig. 4.14 where it is seen that the left and right limits of $g(t)$ are different at t_0; this means that the signal is not continuous at t_0.

Fig. 4.13 Graph of $f(t)$ for Example 4.16

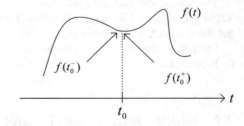

Fig. 4.14 Graph of $g(t)$ for Example 4.17

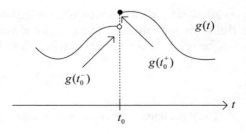

Fig. 4.15 Graph of $f(t)$ for
Example 4.18

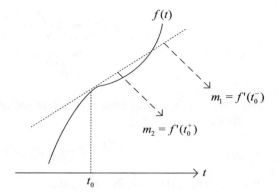

Fig. 4.16 Graph of $f(t)$ for
Example 4.19

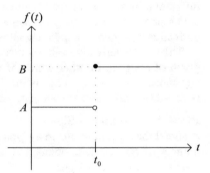

Example 4.18 For the signal shown in Fig. 4.15, it is seen that the left and right
derivative values of the signal are equal to each other at point t_0. That means that the
signal $f(t)$ has a derivative at t_0.

Let us examine a signal that has no derivative at a point.

Example 4.19 Let us examine the derivative of the signal whose graph is shown in
Fig. 4.16 at t_0.

The left derivative value at t_0 is evaluated as

$$f'(t_0^-) = \lim_{h \to 0} \frac{f(t_0 - h) - f(t_0)}{h}$$
$$= \lim_{h \to 0} \frac{A - B}{h}$$

and the right derivative value at t_0 is evaluated as

$$f'\left(t_0^+\right) = \lim_{h \to 0} \frac{f(t_0 + h) - f(t_0)}{-h}$$
$$= \lim_{h \to 0} \frac{B - B}{h}$$
$$= 0.$$

The values of the left derivative and right derivative at t_0 are different from each other, i.e., we have

$$f'\left(t_0^-\right) \neq f'\left(t_0^+\right) \tag{4.46}$$

then the signal $f(t)$ has no derivative at t_0.

Note The continuity of a signal at a point and its derivative at the same point are two different things. A signal may be continuous at a point, but may not have a derivative at that point. We cannot say the same for the opposite case. That is, if a signal has a derivative at a point, then the signal is also continuous at that point.

There is no derivative value at discontinuity points and sharp turning points in the graph of a signal. In other words, if we cannot draw a tangent line to a point, the signal has no derivative at that point. The signal whose graph is depicted in Fig. 4.17 is non-differentiable at points t_0, t_1, and t_2.

Piecewise Continuous Signal
A signal that is continuous in a given interval $[a, b]$ and is not continuous at a finite number of t_1, t_2, ..., t_n points and has one-sided limit at these points is called a piecewise continuous signal.

Fourier Integral
Fourier series are used to express periodic signals in terms of sinusoidal signals. The Fourier integral, on the other hand, is used to express non-periodic, finite signals in terms of sinusoidal signals. The Fourier integral is the limiting case of the Fourier series as the period goes to infinity. That is, the Fourier integral notation is derived from the Fourier series notation. Now let us try to derive the Fourier integral representation of a non-periodic finite signal.

Fig. 4.17 A signal that cannot be differentiated at points t_0, t_1, and t_2

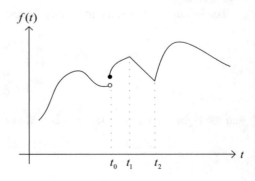

Fourier series representation of the periodic signal $f(t)$ with period T is

$$f(t) = A[0] + \sum_{k=1}^{\infty} (A[k] \cos (kw_0 t) + B[k] \sin (kw_0 t)), \quad w_0 = \frac{2\pi}{T} \tag{4.47}$$

where

$$A[0] = \frac{1}{T} \int_{-\frac{T}{2}}^{\frac{T}{2}} f(t) dt \tag{4.48}$$

$$A[k] = \frac{2}{T} \int_{-\frac{T}{2}}^{\frac{T}{2}} f(t) \cos (kw_0 t) dt \tag{4.49}$$

$$B[k] = \frac{2}{T} \int_{-\frac{T}{2}}^{\frac{T}{2}} f(t) \sin (kw_0 t) dt \tag{4.50}$$

If we substitute (4.48), (4.49), and (4.50) in (4.47) and use the variable v instead of the variable t in the coefficient expressions to avoid confusion, we obtain

$$f(t) = \frac{1}{T} \int_{-\frac{T}{2}}^{\frac{T}{2}} f(v) dv + \frac{2}{T} \sum_{k=1}^{\infty}$$

$$\times \left[\cos (kw_0 t) \int_{-\frac{T}{2}}^{\frac{T}{2}} f(v) \cos (kw_0 v) dv + \sin (kw_0 t) \int_{-\frac{T}{2}}^{\frac{T}{2}} f(v) \sin (kw_0 v) dv \right]$$

where using

$$w_0 = \frac{2\pi}{T} \rightarrow \frac{2}{T} = \frac{w_0}{\pi}$$

we get

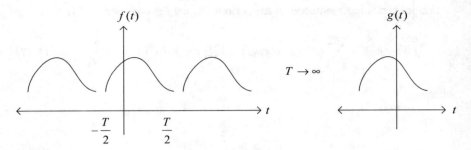

Fig. 4.18 If the period goes to infinity, the resulting signal is equal to one period of the periodic signal around the origin

$$f(t) = \frac{1}{T} \int_{-\frac{T}{2}}^{\frac{T}{2}} f(v)dv + \frac{1}{\pi} \sum_{k=1}^{\infty} \cos(kw_0t)w_0 \int_{-\frac{T}{2}}^{\frac{T}{2}} f(v)\cos(kw_0v)dv$$

$$+ \sin(kw_0t) \int_{-\frac{T}{2}}^{\frac{T}{2}} f(v)\cos(kw_0v)dv$$

$$(4.51)$$

For the case of the period going to infinity i.e., $T \to \infty$, let

$$g(t) = \lim_{T \to \infty} f(t) \qquad (4.52)$$

$g(t)$ equals one period of the signal $f(t)$, and the infinite sum in (4.51) turns into an integral with frontiers 0 and ∞; furthermore, the first expression in (4.51) equals zero, i.e., we have (Fig. 4.18)

$$g(t) = \frac{1}{\pi} \int_0^{\infty} \left(\cos wt \left(\int_{-\infty}^{\infty} g(v)\cos(wv)dv \right) dw + \int_0^{\infty} \sin wt \left(\int_{-\infty}^{\infty} g(v)\cos(wv)dv \right) dv \right)$$

for which if we define the expressions

$$A(w) = \frac{1}{\pi} \int_{-\infty}^{\infty} g(v)\cos(wv)dv \qquad B(w) = \frac{1}{\pi} \int_{-\infty}^{\infty} g(v)\sin(wv)dv$$

and change the parameter v to t, we obtain

$$A(w) = \frac{1}{\pi} \int\limits_{-\infty}^{\infty} g(t) \cos{(wt)} dt \qquad B(w) = \frac{1}{\pi} \int\limits_{-\infty}^{\infty} g(t) \sin{(wt)} dt \qquad (4.53)$$

If we compare (4.53) with

$$A[k] = \frac{2}{T} \int\limits_{-\frac{T}{2}}^{\frac{T}{2}} f(t) \cos{(kw_0 t)} dt \qquad B[k] = \frac{2}{T} \int\limits_{-\frac{T}{2}}^{\frac{T}{2}} f(t) \sin{(kw_0 t)} dt \qquad (4.54)$$

and use the property

$$\int\limits_{t=-\infty}^{\infty} g(t) dt = \int\limits_{t=-\frac{T}{2}}^{\frac{T}{2}} f(t) dt$$

we get

$$A[0] = \frac{\pi}{T} A(0) \quad A[k] = \frac{2\pi}{T} A(w) \Big|_{w=kw_0} \quad B[k] = \frac{2\pi}{T} B(w) \Big|_{w=kw_0} \quad k \in Z^+ \quad (4.55)$$

Then, it is possible to obtain the real Fourier series coefficients of the periodic signal by sampling the Fourier integral expressions $A(w)$ and $B(w)$, which are calculated using only one period of the periodic signal.

Theorem Let $g(t)$ be a piecewise continuous signal defined on an interval, having left and right derivatives at each point of the interval. If

$$\int\limits_{-\infty}^{\infty} |g(t)| dt \qquad (4.56)$$

is a finite number, then $g(t)$ can be expressed with Fourier integral. If there are points at which the signal $g(t)$ is discontinuous, the value of the Fourier integral at a discontinuity point is equal to the average value of the left and right limits of the signal at the discontinuity point.

Fourier integral representation of $g(t)$ in Fig. 4.19 can be achieved using the formulas

Fig. 4.19 Calculation of the
Fourier integral value at a
discontinuity point

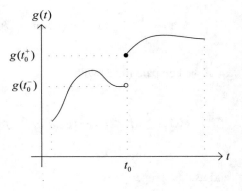

Fig. 4.20 Graph of $f(t)$ for
Example 4.20

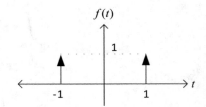

$$g(t) = \int\limits_{0}^{\infty} [A(w)\cos(wt) + B(w)\sin(wt)]dw$$

$$A(w) = \frac{1}{\pi} \int\limits_{-\infty}^{\infty} g(t)\cos(wt)dt \qquad (4.57)$$

$$B(w) = \frac{1}{\pi} \int\limits_{-\infty}^{\infty} g(t)\sin(wt)dt.$$

$g(t)$ is discontinuous at t_0. The value of the signal at the discontinuity point can be
calculated as

$$\int_{0}^{\infty} [A(w)\cos(wt) + B(w)\sin(wt)]dw\big|_{t=t_0} = \frac{g(t_0^-) + g(t_0^+)}{2} \qquad (4.58)$$

Example 4.20 Obtain the Fourier integral representation of $f(t)$ shown in Fig. 4.20.

Solution 4.20 Let us first write a mathematical expression for the given signal and
then use the mathematical expression in Fourier integral equations.

  $f(t)$ can be written in terms of impulses as

$$f(t) = \delta(t+1) + \delta(t-1)$$

which can be used for the calculation of $A(w)$ as

$$A(w) = \frac{1}{\pi} \int_{-\infty}^{\infty} f(t) \cos{(wt)}dt$$

$$= \frac{1}{\pi} \int_{-\infty}^{\infty} (\delta(t+1) + \delta(t-1)) \cos{(wt)}dt$$

$$= \frac{1}{\pi}(\cos{(-w)} + \cos{(w)})$$

$$= \frac{2\cos{(w)}}{\pi}$$

and $B(w)$ can be calculated as

$$B(w) = \frac{1}{\pi} \int_{-\infty}^{\infty} f(t) \sin{(wt)}dt$$

$$= \frac{1}{\pi} \int_{-\infty}^{\infty} (\delta(t+1) + \delta(t-1)) \sin{(wt)}dt$$

$$= \frac{1}{\pi}(\sin{(-w)} + \sin{(w)})$$

$$= 0.$$

Substituting $A(w)$ and $B(w)$ in

$$f(t) = \int_{0}^{\infty} [A(w)\cos{(wt)} + B(w)\sin{(wt)}]dw$$

we obtain

$$f(t) = \frac{2}{\pi} \int_{0}^{\infty} \cos{(w)}\cos{(wt)}dw$$

By equating the definition of the signal with its Fourier integral, we can write the equation

$$\delta(t+1) + \delta(t-1) = \frac{2}{\pi} \int\limits_{0}^{\infty} \cos(w) \cos(wt) dw$$

Example 4.21 Using $f(t)$ in the previous example, we obtain a periodic signal as

$$p(t) = \sum_{k=-\infty}^{\infty} f(t-kT)$$

where T is the period of $p(t)$. Find the Fourier series representation of $p(t)$.

Solution 4.21 Fourier series representation of $p(t)$ is

$$p(t) = A[0] + \sum_{k=1}^{\infty} A[k] \cos\left(k\frac{2\pi}{T}t\right) + \sum_{k=1}^{\infty} B[k] \sin\left(k\frac{2\pi}{T}t\right)$$

where real coefficients can be calculated from Fourier integral as

$$A[0] = \frac{\pi}{T}A(0) \quad A[k] = \frac{2\pi}{T}A(w)\big|_{w=kw_0}$$

$$B[k] = \frac{2\pi}{T}B(w)\big|_{w=kw_0} \quad k = 1,2,\ldots,\infty$$

where using the results

$$A(w) = \frac{2\cos(w)}{\pi} \quad B(w) = 0$$

obtained in the previous example, we get

$$A[0] = \frac{2}{T} \quad A[k] = \frac{4}{T}\cos\left(k\frac{2\pi}{T}\right) \quad B[k] = 0$$

Using the calculated coefficients, we obtain the Fourier series representation of $p(t)$ as

$$p(t) = \frac{2}{T} + \sum_{k=1}^{\infty} \frac{4}{T}\cos\left(k\frac{2\pi}{T}\right)\cos\left(k\frac{2\pi}{T}t\right)$$

Exercise Obtain the Fourier integral representation of the signal $g(t)$ shown in Fig. 4.21.

Note For even signals, we have $B(w) = 0$, and for odd signals, we have $A(w) = 0$.

Fig. 4.21 Graph of $g(t)$ for exercise

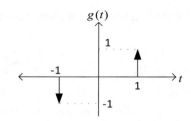

Fig. 4.22 Graph of $f(t)$ for Example 4.22

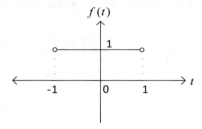

Example 4.22 Consider the Fourier integral representation of the signal depicted in Fig. 4.22.

The signal in Fig. 4.22 can be mathematically expressed as

$$f(t) = \begin{cases} 1 & |t| \leq 1 \\ 0 & \text{otherwise} \end{cases}.$$

Since the signal is even, we have $B(w) = 0$. Although we can directly write that $B(w) = 0$, we will evaluate $B(w)$ mathematically and show that it is equal to zero. If we use the Fourier integral formulas

$$f(t) = \int_0^\infty [A(w)\cos(wt) + B(w)\sin(wt)]dw \qquad (4.59)$$

$$A(w) = \frac{1}{\pi} \int_{-\infty}^{\infty} f(t)\cos(wt)dt \quad B(w) = \frac{1}{\pi} \int_{-\infty}^{\infty} f(t)\sin(wt)dt \qquad (4.60)$$

for the signal, we get the coefficient expressions as

$$A(w) = \frac{1}{\pi} \int_{-1}^{1} 1\cos(wt)dt \qquad B(w) = \frac{1}{\pi} \int_{-1}^{1} 1\sin(wt)dt$$

$$= \frac{\sin(wt)}{\pi w}\Big|_{-1}^{1} \frac{2\sin(w)}{\pi w} \qquad = -\frac{1}{\pi w}\cos(wt)\Big|_{-1}^{1} 0$$

and when these expressions are used in (4.59), we get

$$f(t) = \frac{2}{\pi} \int_0^\infty \frac{\cos{(wt)}\sin{(w)}}{w} dw$$

$f(t)$ is discontinuous at the points $t = -1$ and $t = 1$. The value of the signal's Fourier integral representation at this point is equal to the average of the signal's left and right limit values at the point. That is,

$$\frac{2}{\pi} \int_0^\infty \frac{\cos{(wt)}\sin{(w)}}{w} dw \bigg|_{t=1} = \frac{f(1^-) + f(1^+)}{2} \rightarrow \frac{2}{\pi} \int_0^\infty \frac{\cos{(wt)}\sin{(w)}}{w} dw$$

$$= \frac{1+0}{2} \rightarrow \int_0^\infty \frac{\sin{(2w)}}{w} dw = \frac{\pi}{4}$$

Since the cosine signal is even, the calculation of the above expression at point -1 gives the same integral result. Let us consider the signal in the range $-1 < t < 1$ and calculate the Fourier integral value for the point 0 as in

$$f(0) = \frac{2}{\pi} \int_0^\infty \frac{\cos{(0)}\sin{(w)}}{w} dw \rightarrow 1 = \frac{2}{\pi} \int_0^\infty \frac{\sin{(w)}}{w} dw \rightarrow \int_0^\infty \frac{\sin{(w)}}{w} dw = \frac{\pi}{2}$$

Gathering all the results, it is possible to write a mathematical expression for the signal as

$$\int_0^\infty \frac{\cos{(wt)}\sin{(w)}}{w} dw = \begin{cases} \frac{\pi}{2} & -1 < t < 1 \\ \frac{\pi}{4} & t = 1 \\ 0 & |t| > 1 \end{cases} \tag{4.61}$$

Let us consider the mathematical expression

$$\int_0^\infty \frac{\sin{(w)}}{w} dw = \frac{\pi}{2} \tag{4.62}$$

In (4.62), the upper limit of the integral is infinity. Using (4.62), let us define Si(u) as

$$Si(u) = \int_0^u \frac{\sin{(w)}}{w} dw \tag{4.63}$$

The graphs of

$$\frac{\sin{(w)}}{w} \quad \text{and} \quad \text{Si}(u)$$

are depicted in Fig. 4.23. It is seen from Fig. 4.23 that as u goes to $\pm\infty$, the signal $\text{Si}(u)$ approaches $\pm\frac{\pi}{2}$.

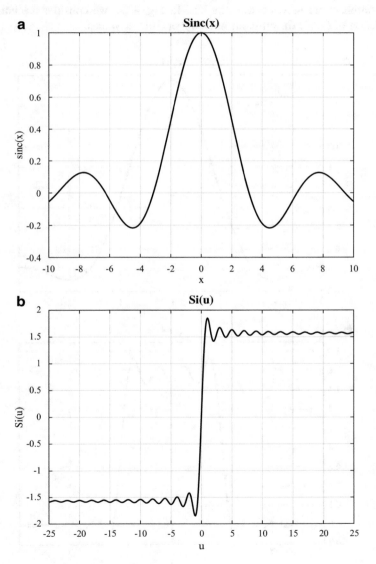

Fig. 4.23 The graphs of $\text{sinc}(w) = \frac{\sin{(w)}}{w}$ and $\text{Si}(u)$

The Fourier integral representation of the square pulse is

$$f(t) = \int_0^\infty \frac{\cos(wt)\sin(w)}{w}\,dw \tag{4.64}$$

where the upper frontier of the integral is ∞. In practical calculations of integrals, finite numbers are used for the frontiers. In Fig. 4.24, we consider the numerical integration of (4.64) for different finite upper frontier values.

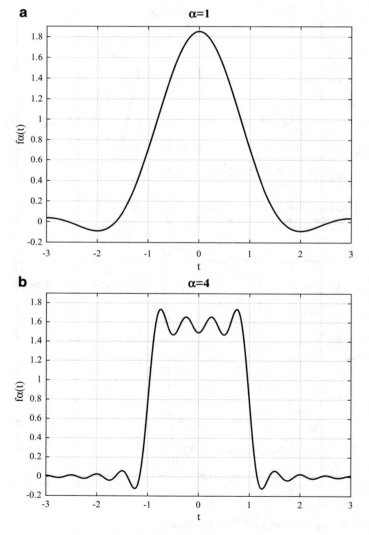

Fig. 4.24 The graphs of $f(t) = \int_0^\alpha \frac{\cos(wt)\sin(w)}{w}\,dw$ for different α values

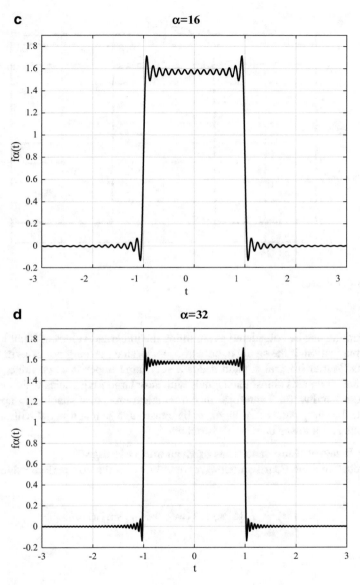

Fig. 4.24 (continued)

It is seen from the graphs in Fig. 4.24 that as the upper limit value of the integral increases, the resulting shape gets closer to the square pulse, and it is also seen that sharpening the corners of the square improves slowly. If we use very large upper frontier values for integral, perfect corners can be achieved.

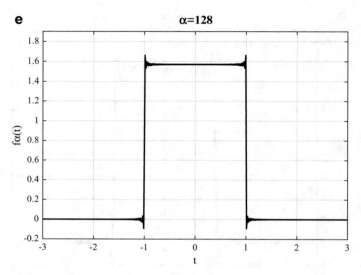

Fig. 4.24 (continued)

Integration can be considered as summing the integrand values at small intervals. In communication, if the signal to be sent contains sharp turning points as in a square wave, its Fourier integral contain needs a very large upper frontier value, and this means that it contains sinusoidal signals with very large frequencies which implies a very large bandwidth. Therefore, in communication, pulse-shaping is applied to eliminate the sharp parts of the signal to be transmitted so that a signal with smoother transitions and a lower bandwidth is obtained.

Fourier Integral Representations of Even and Odd Signals

Let us recall that the Fourier integral representation of the non-periodic signal $f(t)$ is

$$f(t) = \int\limits_{0}^{\infty} [A(w)\cos(wt) + B(w)\sin(wt)]dw \qquad (4.65)$$

where

$$A(w) = \frac{1}{\pi} \int\limits_{-\infty}^{\infty} f(t)\cos(wt)dt \quad B(w) = \frac{1}{\pi} \int\limits_{-\infty}^{\infty} f(t)\sin(wt)dt \qquad (4.66)$$

If $f(t)$ is an even signal, then we have $f(t) = f(-t)$. If we divide the integral interval into two parts in $A(w)$, i.e.,

$$(-\infty \ldots \infty) = (-\infty \ldots 0) \cup (0 \ldots \infty)$$

we get

$$
\begin{aligned}
A(w) &= \frac{1}{\pi} \int_{-\infty}^{\infty} f(t) \cos{(wt)} dt \\
&= \frac{1}{\pi} \left(\int_{-\infty}^{0} f(t) \cos{(wt)} dt + \int_{0}^{\infty} f(t) \cos{(wt)} dt \right) \\
&= \frac{1}{\pi} \left(\int_{0}^{\infty} f(-t) \cos{(-wt)} dt + \int_{0}^{\infty} f(t) \cos{(wt)} dt \right) \\
&= \frac{2}{\pi} \int_{0}^{\infty} f(t) \cos{(wt)} dt.
\end{aligned}
$$

Similarly, for an even signal, $B(w)$ can be calculated as

$$
\begin{aligned}
B(w) &= \frac{1}{\pi} \int_{-\infty}^{\infty} f(t) \sin{(wt)} dt \\
&= \frac{1}{\pi} \left(\int_{-\infty}^{0} f(t) \sin{(wt)} dt + \int_{0}^{\infty} f(t) \sin{(wt)} dt \right) \\
&= \frac{1}{\pi} \left(\int_{0}^{\infty} f(-t) \sin{(-wt)} dt + \int_{0}^{\infty} f(t) \sin{(wt)} dt \right) \\
&= \frac{1}{\pi} \left(\int_{0}^{\infty} -f(t) \sin{(wt)} dt + \int_{0}^{\infty} f(t) \sin{(wt)} dt \right) \\
&= 0.
\end{aligned}
$$

Using the obtained results, we can express the Fourier integral representation of an even signal as

$$f(t) = \int_{0}^{\infty} A(w) \cos{(wt)} dw \tag{4.67}$$

where

$$A(w) = \frac{2}{\pi} \int_{0}^{\infty} f(t) \cos{(wt)} dt \tag{4.68}$$

For even $f(t)$, $B(w)$ equals zero; on the other hand, $A(w)$ is a non-zero expression. Fourier integral representations of odd signals can be derived as

$$f(t) = \int_0^\infty B(w) \sin(wt)dw$$

$$B(w) = \frac{2}{\pi} \int_0^\infty f(t) \sin(wt)dt.$$

(4.69)

For odd signals, unlike even signals, $A(w)$ equals zero, and $B(w)$ is non-zero.

Theorem If $f(t)$ is a right-sided signal, i.e., $f(t) = 0$ if $t < 0$, the Fourier integral representation of $f(t)$ can be either as

$$f(t) = \int_0^\infty A(w) \cos(wt)dw$$

(4.70)

where

$$A(w) = \frac{2}{\pi} \int_0^\infty f(t) \cos(wt)dt$$

(4.71)

or as

$$f(t) = \int_0^\infty B(w) \sin(wt)dw$$

(4.72)

where

$$B(w) = \frac{2}{\pi} \int_0^\infty f(t) \sin(wt)dt$$

(4.73)

Proof A right-sided signal $f(t)$ takes values only for $t > 0$, and it is always equal to zero for $t < 0$. Let us define $g(t)$ as

$$g(t) = f(t) + f(-t)$$

which is an even signal, i.e., $g(t) = g(-t)$, and the Fourier integral representation of $g(t)$ can be written as

$$g(t) = \int_0^\infty A(w) \cos{(wt)}dw \tag{4.74}$$

where

$$A(w) = \frac{2}{\pi} \int_0^\infty g(t) \cos{(wt)}dt \tag{4.75}$$

In (4.75), the integral is evaluated from 0 to ∞, i.e., it is calculated for positive t values. Since $f(t)$ is a right-sided signal, the expression

$$0 \leq t < \infty \quad g(t) = f(t) + f(-t) \tag{4.76}$$

can be written as

$$0 \leq t < \infty \quad g(t) = f(t) \tag{4.77}$$

since for positive t values we have $f(-t) = 0$. Substituting (4.77) in (4.75), we obtain

$$A(w) = \frac{2}{\pi} \int_0^\infty f(t) \cos{(wt)}dt \tag{4.78}$$

and for $t > 0$ we have

$$g(t) = f(t) \rightarrow g(t) = \int_0^\infty A(w) \cos{(wt)}dt \tag{4.79}$$

Similarly, using $f(t)$ we can obtain the odd signal $g(t)$, and considering the Fourier integral representation of $g(t)$, we can prove the second part of the theorem. To summarize, a right-sided signal can be illustrated by Fourier integral representations of both even and odd signals.

Exercise   Let $f(t)$ be a left-sided signal such that $f(t) = 0$ if $t > 0$ and $f(t) \neq 0$ if $t < 0$. Obtain sine and cosine Fourier integral representations of $f(t)$.

Example 4.23   For $f(t) = e^{-kt}u(t)$, $k > 0$ where $u(t)$ is the unit step signal, obtain sine and cosine Fourier integral representations of $f(t)$.

Solution 4.23 If $f(t) = 0$ for $t < 0$, then $f(t)$ is a right-sided signal. Such a signal has both sine and cosine Fourier integral representations. Cosine Fourier integral representation can be obtained using

$$f(t) = \int_0^\infty A(w) \cos(wt) dw \tag{4.80}$$

where

$$A(w) = \frac{2}{\pi} \int_0^\infty f(t) \cos(wt) dt$$

For the evaluation of the integral expression in

$$A(w) = \frac{2}{\pi} \int_0^\infty e^{-kt} \cos(wt) dt$$

we can employ the partial integration method

$$\int u\, dv = uv - \int v\, du$$

where using $u = e^{-kt}$ and $dv = \cos(wt) dt$, we get

$$\int_0^\infty e^{-kt} \cos(wt) dt = \left(\frac{e^{-kt}}{w} \sin(wt) \Big|_0^\infty \right) - \int_0^\infty \frac{1}{w} \sin(wt)(-k) e^{-kt} dt$$

$$= 0 - 0 + \frac{k}{w} \int_0^\infty \sin(wt) t e^{-kt} dt.$$

from which we obtain

$$\int_0^\infty e^{-kt} \cos(wt) dt = \frac{k}{w} \int_0^\infty \sin(wt) e^{-kt} dt \tag{4.81}$$

Let us use the partial integration evaluation method for the right-hand side of (4.81); for this purpose let $u = e^{-kt}$ and $dv = \sin(wt) dt$; and then we have

$$\frac{k}{w} \int_0^\infty e^{-kt} \sin{(wt)}dt \rightarrow$$

$$\rightarrow \frac{k}{w} \left[\left(-\frac{e^{-kt}}{w} \cos{(wt)} \Big|_0^\infty \right) - \int_0^\infty -\frac{\cos{(wt)}}{w}(-k)e^{-kt}dt \right]$$

$$\rightarrow \frac{k}{w} \left[\frac{1}{w} - \frac{k}{w} \int_0^\infty e^{-kt} \cos{(wt)}dt \right]$$

Thus, we have

$$\frac{k}{w} \int_0^\infty e^{-kt} \sin{(wt)}dt = \frac{k}{w} \left[\frac{1}{w} - \frac{k}{w} \int_0^\infty e^{-kt} \cos{(wt)}dt \right] \qquad (4.82)$$

Using (4.82) on the right-hand side of (4.81), we can write (4.81) as in

$$\int_0^\infty e^{-kt} \cos{(wt)}dt = \frac{k}{w^2} - \frac{k^2}{w^2} \int_0^\infty e^{-kt} \cos{(wt)}dt \qquad (4.83)$$

from which we obtain

$$\int_0^\infty e^{-kt} \cos{(wt)}dt = \frac{k}{k^2 + w^2} \qquad (4.84)$$

Substituting (4.84) in

$$A(w) = \frac{2}{\pi} \int_0^\infty e^{-kt} \cos{(wt)}dt$$

we get

$$A(w) = \frac{1}{\pi} \frac{2k}{k^2 + w^2}$$

Using $A(w)$ in Fourier integral representation of $f(t)$ in (4.80), we obtain

$$f(t) = \frac{2k}{\pi} \int_0^\infty \frac{\cos{(wt)}}{k^2 + w^2} dw \, t > 0$$

If we equate the Fourier integral representation of $f(t)$ to the definition of $f(t)$

$$f(t) = e^{-kt} \quad t > 0, \quad k > 0$$

we get

$$e^{-kt} = \frac{2k}{\pi} \int_{0}^{\infty} \frac{\cos(wt)}{k^2 + w^2} dw \, t > 0, \quad k > 0$$

which can be written as

$$\int_{0}^{\infty} \frac{\cos(wt)}{k^2 + w^2} dw = \frac{\pi}{2k} e^{-kt} \, t > 0, \quad k > 0$$

which is called Laplace integral.

4.2.1   Fourier Cosine and Sine Transforms

For a right-sided signal $f(t)$, the cosine representation is

$$f(t) = \int_{0}^{\infty} A(w) \cos(wt) dw \tag{4.85}$$

where

$$A(w) = \frac{2}{\pi} \int_{0}^{\infty} f(t) \cos(wt) dt \tag{4.86}$$

and the sine representation of $f(t)$ can be obtained using

$$f(t) = \int_{0}^{\infty} B(w) \sin(wt) dw \tag{4.87}$$

where

$$B(w) = \frac{2}{\pi} \int_{0}^{\infty} f(t) \sin(wt) dt \tag{4.88}$$

Using $A(w)$, let us define

$$\widehat{f}_c(w) = \sqrt{\frac{\pi}{2}}A(w) \tag{4.89}$$

from which we obtain

$$\widehat{f}_c(w) = \sqrt{\frac{2}{\pi}}\int_0^\infty f(t)\cos{(wt)}dt$$

$$f(t) = \sqrt{\frac{2}{\pi}}\int_0^\infty \widehat{f}_c(w)\cos{(wt)}dw \tag{4.90}$$

where $\widehat{f}_c(w)$ is the cosine transform of $f(t)$. Obtaining $f(t)$ from $\widehat{f}_c(w)$ is called inverse cosine transform.

In a similar manner using $B(w)$, let us define

$$\widehat{f}_s(w) = \sqrt{\frac{\pi}{2}}B(w) \tag{4.91}$$

where substituting (4.73), we get

$$\widehat{f}_s(w) = \sqrt{\frac{2}{\pi}}\int_0^\infty f(t)\sin{(wt)}dt \tag{4.92}$$

It is possible to obtain the signal itself using the sine transform of $f(t)$. This operation can be performed using the inverse sine transform. The expression for the inverse sine transform is

$$f(t) = \sqrt{\frac{2}{\pi}}\int_0^\infty \widehat{f}_s(w)\sin{(wt)}dw \tag{4.93}$$

Example 4.24 Find the sine and cosine transforms of $f(t)$ shown in Fig. 4.25.

Solution 4.24 We can calculate the cosine transform of the signal as

$$\widehat{f}_c(w) = \sqrt{\frac{2}{\pi}}\int_0^\infty f(t)\cos{(wt)}dt \rightarrow \widehat{f}_c(w) = \sqrt{\frac{2}{\pi}}\int_0^a k\cos{(wt)}dt$$

$$= \sqrt{\frac{2}{\pi}}\frac{k}{w}\left(\sin{(wt)}\Big|_0^a \right) = \sqrt{\frac{2}{\pi}}k\frac{\sin{(aw)}}{w}.$$

The sine transform of the signal can be calculated as

Fig. 4.25 Graph of $f(t)$ for
Example 4.24

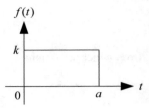

$$\widehat{f}_s(w) = \sqrt{\frac{2}{\pi}} \int_0^{\infty} f(t) \sin{(wt)}dt \rightarrow \widehat{f}_s(w) = \sqrt{\frac{2}{\pi}} \int_0^{a} k \sin{(wt)}dt$$

$$= \sqrt{\frac{2}{\pi}}k \left(\frac{1 - \cos{(aw)}}{w} \right).$$

We can get some interesting results by using these transformations in the inverse transformation formulas. Using the inverse cosine transform formula, we get

$$f(t) = \sqrt{\frac{2}{\pi}} \int_0^{\infty} \widehat{f}_c(w) \cos{(wt)}dw$$

$$= \sqrt{\frac{2}{\pi}} \int_0^{\infty} \sqrt{\frac{2}{\pi}}k \frac{\sin{(aw)}}{w} \cos{(wt)}dw$$

from which we obtain

$$f(t) = \frac{2k}{\pi} \int_0^{\infty} \frac{\sin{(aw)}}{w} \cos{(wt)}dw \qquad (4.94)$$

On the interval $0 \leq t < a$, we have $f(t) = k$, and if this result is used in (4.94), we get

$$k = \frac{2k}{\pi} \int_0^{\infty} \frac{\sin{(aw)}}{w} \cos{(wt)}dw \rightarrow \frac{\pi}{2} = \int_0^{\infty} \frac{\sin{(aw)}}{w} \cos{(wt)}dw$$

That is

$$\int_0^{\infty} \frac{\sin{(aw)}}{w} \cos{(wt)}dw = \frac{\pi}{2}, \quad 0 \leq t < a \qquad (4.95)$$

4.2.2   Fourier Transform

In this section, we first explain the Fourier transforms of non-periodic signals. Then, we consider the Fourier transforms of periodic signals. We obtain Fourier transform and inverse Fourier transform expressions of non-periodic signals from the Fourier integral formulas. The Fourier integral formula is used for non-periodic signals. Fourier integral representation for any non-periodic signal $f(t)$ is

$$f(t) = \int_0^\infty (A(w)\cos(wt) + B(w)\sin(wt))dw \qquad (4.96)$$

where we have

$$A(w) = \frac{1}{\pi} \int_{-\infty}^\infty f(t)\cos(wt)dt \quad B(w) = \frac{1}{\pi} \int_{-\infty}^\infty f(t)\sin(wt)dt \qquad (4.97)$$

Let us combine $A(w)$ and $B(w)$ in one expression as

$$\widehat{f}(w) = \sqrt{\frac{\pi}{2}}(A(w) - jB(w)) \qquad (4.98)$$

where $\widehat{f}(w)$ is called the Fourier transform of the signal $f(t)$. Substituting (4.97) in (4.98), we get

$$\widehat{f}(w) = \sqrt{\frac{\pi}{2}}(A(w) - jB(w))$$

$$= \sqrt{\frac{\pi}{2}}\left(\frac{1}{\pi} \int_{-\infty}^\infty f(t)\cos(wt)dt - \frac{j}{\pi} \int_{-\infty}^\infty f(t)\sin(wt)dt\right)$$

$$= \frac{1}{\sqrt{2\pi}}\left(\int_{-\infty}^\infty f(t)[\cos(wt) - j\sin(wt)]dt\right)$$

$$= \frac{1}{\sqrt{2\pi}} \int_{-\infty}^\infty f(t)e^{-jwt}dt.$$

Thus, Fourier transform formula for $f(t)$ is obtained as

$$\widehat{f}(w) = \frac{1}{\sqrt{2\pi}} \int_{-\infty}^\infty f(t)e^{-jwt}dt \qquad (4.99)$$

where the coefficient $\frac{1}{\sqrt{2\pi}}$ is not used by some authors and Fourier transform formula can also be defined as

$$\widehat{f}(w) = \int_{-\infty}^{\infty} f(t)e^{-jwt}dt \qquad (4.100)$$

which corresponds to $\widehat{f}(w) = \pi(A(w) - jB(w))$.

$f(t)$ can be obtained from $\widehat{f}(w)$. To achieve this goal, first, let us multiply $\widehat{f}(w)$ by e^{jwt} as in

$$\widehat{f}(w)e^{jwt} = \sqrt{\frac{\pi}{2}}(A(w) - jB(w))e^{jwt} \qquad (4.101)$$

where using the identity $e^{jwt} = \cos(wt) + j\sin(wt)$ and evaluating the integral of both sides of (4.101), we obtain

$$\int_{w=-\infty}^{\infty} \widehat{f}(w)e^{jwt}dw = \sqrt{\frac{\pi}{2}} \int_{w=-\infty}^{\infty} (A(w) - jB(w))(\cos(wt) + j\sin(wt))dw$$

$$(4.102)$$

On the right side of (4.102), $A(w)$ and $\cos(wt)$ are even signals, and $B(w)$ and $\sin(wt)$ are odd signals. When an even signal is multiplied by an odd signal, the resulting signal becomes an odd signal. The integral of an odd signal is zero. For instance, the product $A(w)\sin(wt)$ is an odd signal, and we have

$$\int_{w=-\infty}^{\infty} A(w)\sin(wt)dw = 0 \qquad (4.103)$$

Expanding the multiplications on the right-hand side of (4.102), and using (4.103), we obtain

$$\int_{w=-\infty}^{\infty} \widehat{f}(w)e^{jwt}dw = \sqrt{\frac{\pi}{2}} \int_{w=-\infty}^{\infty} [A(w)\cos(wt) + B(w)\sin(wt)]dw \qquad (4.104)$$

The right-hand side of (4.94) equals $2f(t)$ where the coefficient 2 is due to the integral frontiers ranging from $-\infty$ to $+\infty$. In (4.96), frontiers range from 0 to $+\infty$. Thus, from (4.104) we obtain

$$\int_{w=-\infty}^{\infty} \widehat{f}(w)e^{jwt}dw = 2\sqrt{\frac{\pi}{2}}f(t) \qquad (4.105)$$

from which $f(t)$ can be written as

$$f(t) = \frac{1}{\sqrt{2\pi}} \int\limits_{w=-\infty}^{\infty} \widehat{f}(w)e^{jwt}\,dw \qquad (4.106)$$

which is called inverse Fourier transform. To summarize, if the expression

$$\widehat{f}(w) = \frac{1}{\sqrt{2\pi}} \int\limits_{t=-\infty}^{\infty} f(t)e^{-jwt}\,dt \qquad (4.107)$$

is used for the Fourier transform, then the inverse Fourier transform is calculated using

$$f(t) = \frac{1}{\sqrt{2\pi}} \int\limits_{w=-\infty}^{\infty} \widehat{f}(w)e^{jwt}\,dw \qquad (4.108)$$

On the other hand, if the Fourier transform is defined as

$$\widehat{f}(w) = \int\limits_{t=-\infty}^{\infty} f(t)e^{-jwt}\,dt \qquad (4.109)$$

then the inverse Fourier transform is calculated using

$$f(t) = \frac{1}{\sqrt{2\pi}} \int\limits_{w=-\infty}^{\infty} \widehat{f}(w)e^{jwt}\,dw \qquad (4.110)$$

The notation we will use throughout the book for the Fourier transform and the inverse Fourier transform is

$$\widehat{f}(w) = \mathrm{FT}(f(t)) \quad f(t) = \mathrm{FT}^{-1}\left(\widehat{f}(w)\right)$$

which can also be simplified as

$$\widehat{f} = \mathrm{FT}(f) \quad f = \mathrm{FT}^{-1}\left(\widehat{f}\right)$$

The parameter w in the Fourier transform represents the angular frequency and is equal to $2\pi f$, i.e., $w = 2\pi f$, where f is the frequency in Hertz.

The Fourier transform can also be expressed in terms of f as

$$\widehat{f}(w) = \frac{1}{\sqrt{2\pi}} \int\limits_{t=-\infty}^{\infty} f(t)e^{-jwt}dt \rightarrow$$

$$\widehat{f}(2\pi f) = \frac{1}{\sqrt{2\pi}} \int\limits_{t=-\infty}^{\infty} f(t)e^{-j2\pi ft}dt \rightarrow \widehat{f}(f) = \frac{1}{\sqrt{2\pi}} \int\limits_{t=-\infty}^{\infty} f(t)e^{-j2\pi ft}dt$$

And similarly, we can express inverse Fourier transform using the parameter f as in

$$f(t) = \frac{1}{\sqrt{2\pi}} \int\limits_{w=-\infty}^{\infty} \widehat{f}(2\pi f)e^{j2\pi ft}2\pi df \rightarrow f(t) = \sqrt{2\pi} \int\limits_{w=-\infty}^{\infty} \widehat{f}(f)e^{j2\pi ft}df$$

Thus, Fourier transform and its inverse can be evaluated using

$$\widehat{f}(f) = \frac{1}{\sqrt{2\pi}} \int\limits_{t=-\infty}^{\infty} f(t)e^{-j2\pi ft}dt \quad \text{and} \quad f(t) = \sqrt{2\pi} \int\limits_{w=-\infty}^{\infty} \widehat{f}(f)e^{j2\pi ft}df \quad (4.111)$$

Note In general, if Fourier transform of $f(t)$ is defined as

$$\widehat{f}(w) = c_1 \int\limits_{-\infty}^{\infty} f(t)e^{-jwt}dt \qquad (4.112)$$

then the inverse Fourier transform is calculated using

$$f(t) = c_2 \int\limits_{-\infty}^{\infty} \widehat{f}(w)e^{-jwt}dt \qquad (4.113)$$

and c_1 and c_2 satisfy

$$c_1 \times c_2 = \frac{1}{2\pi} \qquad (4.114)$$

That is, we can choose the coefficients c_1 and c_2 as

$$c_1 = 1 \quad c_2 = \frac{1}{2\pi}$$

or as

$$c_1 = \frac{1}{\sqrt{2\pi}} \quad c_2 = \frac{1}{\sqrt{2\pi}}$$

or as

$$c_1 = \frac{1}{2\pi} \quad c_2 = 1$$

The general expressions for Fourier transform and inverse Fourier transform involving parameter f can be defined as

$$\widehat{f}(f) = c_1 \int_{t=-\infty}^{\infty} f(t)e^{-j2\pi ft} dt \quad f(t) = c_2 \int_{w=-\infty}^{\infty} \widehat{f}(f)e^{j2\pi ft} df \qquad (4.115)$$

where c_1 and c_2 satisfy

$$c_1 \times c_2 = 1 \qquad (4.116)$$

The Relationship Between the Fourier Series Coefficients of a Periodic Signal and the Fourier Transform of One Period of the Periodic Signal

Assume that $f(t)$ is a non-periodic signal, and using $f(t)$ the periodic signal $g(t)$ is obtained as

$$g(t) = \sum_{m=-\infty}^{\infty} f(t - mT) \qquad (4.117)$$

where T is the period of $g(t)$, and T is selected in a way that shifted replicas of $f(t)$ do not overlap each other. In this case, one period of $g(t)$ equals $f(t)$.

The Fourier series representation of $g(t)$ is

$$g(t) = \sum_{k=-\infty}^{\infty} G[k]e^{jk\frac{2\pi}{T}t} \qquad (4.118)$$

where

$$G[k] = \frac{1}{T} \int_T g(t)e^{-jk\frac{2\pi}{T}} dt \qquad (4.119)$$

The Fourier transform of $f(t)$ is

$$\widehat{f}(w) = \frac{1}{\sqrt{2\pi}} \int_{t=-\infty}^{\infty} f(t)e^{-jwt} dt \qquad (4.120)$$

The relationship between (4.119) and (4.120) is described as

$$G[k] = \left(\frac{\sqrt{2\pi}}{T} \widehat{f}(w) \right) \Bigg|_{w = kw_0} \qquad w_0 = \frac{2\pi}{T} \qquad (4.121)$$

Example 4.25 Find the Fourier transform of the signal shown in Fig. 4.26.

Solution 4.25 Using the Fourier transform formula in (4.120), we can calculate the Fourier transform of the given signal as

$$
\begin{aligned}
\widehat{f}(w) &= \frac{1}{\sqrt{2\pi}} \int_{t=-\infty}^{\infty} f(t) e^{-jwt} dt \\
&= \frac{1}{\sqrt{2\pi}} \int_{t=-1}^{1} 1 \times e^{-jwt} dt \\
&= \frac{1}{\sqrt{2\pi}} \frac{1}{(-jw)} \left(e^{-jwt} \big|_{-1}^{1} \right) \\
&= \frac{1}{\sqrt{2\pi}} \frac{1}{(jw)} \underbrace{\left(e^{jw} - e^{-jw} \right)}_{= 2j \sin(w)} \\
&= \frac{1}{\sqrt{2\pi}} \frac{1}{(jw)} 2j \sin(w) \\
&= \sqrt{\frac{2}{\pi}} \frac{\sin(w)}{w}
\end{aligned}
$$

where the result contains

$$\frac{\sin(w)}{w}$$

sinc(t) signal is used in digital signal processing and mathematics. In signal processing, sinc(t) is defined as

$$\text{sinc}(t) = \frac{\sin(\pi t)}{\pi t} \qquad (4.122)$$

whereas in mathematics sinc(t) is defined as

Fig. 4.26 Graph of $f(t)$ for Example 4.26

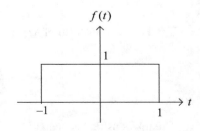

$$\text{sinc}(t) = \frac{\sin{(t)}}{t} \tag{4.123}$$

In this book, we will use the definition of sinc(t) used in signal processing.
   The Fourier transform

$$\widehat{f}(w) = \sqrt{\frac{2}{\pi}} \frac{\sin{(w)}}{w} \tag{4.124}$$

can be expressed in terms of

$$\text{sinc}(\cdot)$$

as

$$\widehat{f}(w) = \sqrt{\frac{2}{\pi}} \text{sinc}\left(\frac{w}{\pi}\right) \tag{4.125}$$

   The Fourier transform integral may not be computable for every non-periodic signal. A signal must satisfy a number of conditions for its Fourier transform integral to be computable.

Theorem Let $f(t)$ be a non-periodic signal. In order for $f(t)$ to have Fourier transform, it must satisfy the following conditions:

1. The absolute value of $f(t)$ must be integrable with respect to the time parameter.
2. $f(t)$ must be a piecewise continuous signal over every finite interval.

Example 4.26 Calculate the Fourier transform, i.e., $\widehat{f}(w)$, of the signal shown in Fig. 4.27.

Solution 4.26 The signal shown in Fig. 4.27 can be written in terms of the impulse signals as

$$f(t) = \delta(t + \pi) + \delta(t - \pi) \tag{4.126}$$

The Fourier transform of (4.126) can be computed as in

Fig. 4.27 Graph of $f(t)$ for Example 4.26

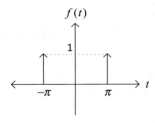

$$\widehat{f}(w) = \frac{1}{\sqrt{2\pi}} \int_{t=-\infty}^{\infty} f(t)e^{-jwt}dt$$

$$= \frac{1}{\sqrt{2\pi}} \int_{t=-\infty}^{\infty} (\delta(t+\pi) + \delta(t-\pi))e^{-jwt}dt$$

$$= \frac{1}{\sqrt{2\pi}} \left(e^{-jw(-\pi)} + e^{-jw\pi} \right)$$

$$= \frac{2\cos(\pi w)}{\sqrt{2\pi}}$$

$$= \sqrt{\frac{2}{\pi}}\cos(\pi w)$$

where we used the property

$$\int_{t=-\infty}^{\infty} f(t)\delta(t-t_0)dt = f(t_0) \tag{4.127}$$

for the calculation of the integral.

The inverse Fourier transform can be calculated as in

$$f(t) = \frac{1}{\sqrt{2\pi}} \int_{w=-\infty}^{\infty} \widehat{f}(w)e^{jwt}dw$$

$$= \frac{1}{\sqrt{2\pi}} \int_{w=-\infty}^{\infty} \sqrt{\frac{2}{\pi}}\cos(\pi w)e^{jwt}dw$$

resulting in

$$f(t) = \frac{1}{\pi} \int_{w=-\infty}^{\infty} \cos(\pi w)e^{jwt}dw \tag{4.128}$$

Equating the definition of $f(t)$ in (4.126), (4.127) and (4.128), we obtain

$$\delta(t+\pi) + \delta(t-\pi) = \frac{1}{\pi} \int_{w=-\infty}^{\infty} \cos(\pi w)e^{jwt}dw \tag{4.129}$$

Fig. 4.28 Graph of $f(t)$ for Example 4.27

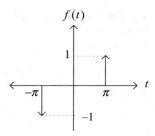

$f(t)$

Example 4.27 Calculate the Fourier transform, i.e., $\widehat{f}(w)$, of the signal shown in Fig. 4.28.

Solution 4.27 This problem is similar to the previous one we solved; both questions contain two fundamental signals that are very important in communication engineering. The signal shown in Fig. 4.28 can be written in terms of impulse signals as

$$f(t) = -\delta(t + \pi) + \delta(t - \pi) \tag{4.130}$$

Fourier transform of (4.130) can be evaluated as

$$
\begin{aligned}
\widehat{f}(w) &= \frac{1}{\sqrt{2\pi}} \int_{t=-\infty}^{\infty} f(t)e^{-jwt}dt \\
&= \frac{1}{\sqrt{2\pi}} \int_{t=-\infty}^{\infty} (-\delta(t + \pi) + \delta(t - \pi))e^{-jwt}dt \\
&= \frac{1}{\sqrt{2\pi}} \left(-e^{-jw(-\pi)} + e^{-jw\pi} \right) \\
&= -\frac{2j\sin(\pi w)}{\sqrt{2\pi}} \\
&= -j\sqrt{\frac{2}{\pi}}\sin(\pi w).
\end{aligned}
\tag{4.131}
$$

In $\widehat{f}(w)$ calculation, we used the properties

$$\int_{t=-\infty}^{\infty} f(t)\delta(t - t_0)dt = f(t_0) \tag{4.132}$$

and

Fig. 4.29 Graph of $\widehat{f}(w)$ for
Example 4.28

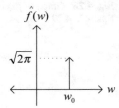

$$\sin\left(\theta\right)=\frac{1}{2j}\left(e^{j\theta}-e^{-j\theta}\right) \tag{4.133}$$

at the second and third lines of (4.131). The result of these two examples should be well kept in mind. The Fourier transform of two impulses in the same direction gives the cosine signal, while the Fourier transform of two impulse signals with opposite directions gives the sine signal.

Example 4.28 The Fourier transform of a continuous signal is depicted in Fig. 4.29. Find the time domain expression of this signal, i.e., $f(t)=?$

Solution 4.28 The signal in Fig. 4.29 can be written mathematically as

$$\widehat{f}(w)=\sqrt{2\pi}\delta(w-w_0) \tag{4.134}$$

The Fourier transform of (4.134) can be calculated as

$$f(t)=\frac{1}{\sqrt{2\pi}}\int\limits_{w=-\infty}^{\infty}\widehat{f}(w)e^{jwt}dw$$

$$=\frac{1}{\sqrt{2\pi}}\int\limits_{w=-\infty}^{\infty}\sqrt{2\pi}\delta(w-w_0)e^{jwt}dw$$

where using

$$\int\delta(w-w_0)g(w)dw=g(w_0)$$

we obtain

$$f(t)=e^{jw_0t}$$

Using the result we obtained, we can write the pair

$$e^{jw_0t} \overset{\text{FT}}{\leftrightarrow} \sqrt{2\pi}\delta(w - w_0) \tag{4.135}$$

Property The Fourier transform is a linear operation. In other words, let a, b be real numbers, and let FT $(\cdot)$ denote the Fourier transform; then the Fourier transform of linear combinations of signals $f(t)$, $g(t)$ satisfies

$$\text{FT}(af(t) + bg(t)) = a\text{FT}(f(t)) + b\text{FT}(g(t)) \tag{4.136}$$

Example 4.29 Compute the Fourier transform of $f(t) = \cos(w_0t)$, i.e., $\widehat{f}(w) = ?$

Solution 4.29 The Fourier transform of the given signal can be calculated as in

$$
\begin{aligned}
f(t) &= \cos(w_0t) \\
&= \frac{1}{2}\left(e^{jw_0t} + e^{-jw_0t}\right)
\end{aligned}
\qquad
\begin{aligned}
\text{FT}(f(t)) &= \frac{1}{2}\left(\text{FT}\left(e^{jw_0t}\right) + \text{FT}\left(e^{-jw_0t}\right)\right) \\
&= \frac{1}{2}\left[\sqrt{2\pi}\delta(w - w_0) + \sqrt{2\pi}\delta(w + w_0)\right] \\
&= \sqrt{\frac{\pi}{2}}[\delta(w - w_0) + \delta(w + w_0)]
\end{aligned}
$$

The graph of the signal resulting from the Fourier transform is depicted in Fig. 4.30.

It is seen from Fig. 4.30 that the Fourier transform of the cosine signal includes two impulse signals, which means that there is only one kind of frequency within the cosine signal, which is w_0.

Let us rewrite this result we have obtained in the form of transformation pair as in

$$\cos(w_0t) \overset{\text{FT}}{\leftrightarrow} \sqrt{\frac{\pi}{2}}[\delta(w - w_0) + \delta(w + w_0)] \qquad w_0 = \frac{2\pi}{T} \tag{4.137}$$

where T is the period of cosine signal.

Fig. 4.30 The graph of the Fourier transform for Example 4.29

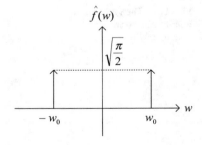

Example 4.30 Find the Fourier transform of $f(t) = \sin(w_0 t)$, i.e., $\widehat{f}(w) = ?$

Solution 4.30 The Fourier transform of the given signal can be calculated as

$$
\begin{aligned}
f(t) &= \sin(w_0 t) \\
&= \frac{1}{2j}\left(e^{jw_0 t} - e^{-jw_0 t}\right)
\end{aligned}
\qquad
\begin{aligned}
\mathrm{FT}(f(t)) &= \frac{1}{2j}\left(\mathrm{FT}\left(e^{jw_0 t}\right) - \mathrm{FT}\left(e^{-jw_0 t}\right)\right) \\
&= \frac{1}{2j}\left[\sqrt{2\pi}\delta(w - w_0) - \sqrt{2\pi}\delta(w + w_0)\right] \\
&= j\sqrt{\frac{\pi}{2}}[\delta(w + w_0) - \delta(w - w_0)]
\end{aligned}
$$

where it is seen that $\widehat{f}(w)$ is a complex signal. The mathematical expression of $j\widehat{f}(w)$ can be written as

$$
j\widehat{f}(w) = \sqrt{\frac{\pi}{2}}[\delta(w - w_0) - \delta(w + w_0)] \tag{4.138}
$$

whose graph is shown in Fig. 4.31.

We can express the result we obtained in the form of transformation pair as

$$
j\sin(w_0 t) \overset{\mathrm{FT}}{\longleftrightarrow} \sqrt{\frac{\pi}{2}}[\delta(w - w_0) - \delta(w + w_0)] \qquad w_0 = \frac{2\pi}{T} \tag{4.139}
$$

Property Assume that $f(t)$ is a periodic signal with period T. Let us define

$$
g(t) = \frac{d^n f(t)}{dt^n}
$$

The relationship between the Fourier series coefficients of $g(t)$ and $f(t)$ can be written as

Fig. 4.31 The graph of the Fourier transform for Example 4.30

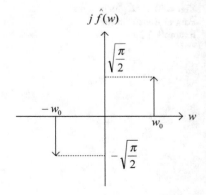

$$f(t) = \sum_{k=-\infty}^{\infty} F[k]e^{jkw_0t} \qquad g(t) = \sum_{k=-\infty}^{\infty} G[k]e^{jkw_0t} \qquad w_0 = \frac{2\pi}{T} \qquad (4.140)$$

where

$$G[k] = (jkw_0)^n F[k].$$

Example 4.31 A single period of a periodic signal is shown in Fig. 4.32. The period of the signal is T. Find the Fourier series representation of this signal.

Solution 4.31 We can find the Fourier series representation of $f(t)$ using the Fourier series representation formula, or we can first take the first-order derivative of the signal, calculate the Fourier series coefficients of the derivative signal, and then find the Fourier series coefficients of the $f(t)$.

Let

$$g(t) = \dot{f}(t)$$

To find $g(t)$, let us first write $f(t)$ in terms of unit step signals and then take its derivative as in

$$f(t) = u(t+1) - u(t-1) \qquad g(t) = \delta(t+1) - \delta(t-1)$$

Both $g(t)$ and $f(t)$ have the same period. The graph of $g(t)$ is depicted in Fig. 4.33.

Fig. 4.32 Graph of $f(t)$ for Example 4.31

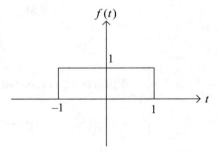

Fig. 4.33 The graph of $\delta(t+1) - \delta(t-1)$

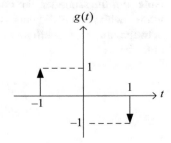

The relationship between the Fourier series coefficients of $g(t)$, i.e., $G[k]$, and the Fourier series coefficients of the $f(t)$, $F[k]$, is as

$$G[k] = (jkw_0)F[k] \qquad w_0 = \frac{2\pi}{T} \qquad\qquad (4.141)$$

The Fourier series coefficients of the signal $g(t)$ can be calculated as

$$
\begin{aligned}
G[k] &= \frac{1}{T}\int_T g(t)e^{-jkw_0 t}\,dt \\
&= \frac{1}{T}\int_T [\delta(t+1) - \delta(t-1)]e^{-jkw_0 t}\,dt \\
&= \frac{1}{T}\left[e^{-jkw_0(-1)} - e^{-jkw_0} \right] \\
&= \frac{1}{T}2j\sin(kw_0) \\
&= \frac{2j}{T}\sin(kw_0).
\end{aligned}
$$

Using (4.141), we can calculate the Fourier series coefficients of $f(t)$ as

$$G[k] = (jkw_0)F[k] \quad w_0 = \frac{2\pi}{T}$$

from which we get

$$
\begin{aligned}
F[k] &= \frac{1}{jk\dfrac{2\pi}{T}}\frac{2j}{T}\sin(kw_0) \\
&= \frac{1}{k\pi}\sin(kw_0).
\end{aligned}
\qquad\qquad (4.142)
$$

For $T = 4$, (4.142) can be evaluated as

$$F[k] = \frac{1}{k\pi}\sin\left(\frac{k\pi}{2}\right)$$

Note For the reminder, the Fourier series representation of a periodic signal can be written with real coefficients as well as using complex coefficients. The relationship between the real coefficients $A[k]$, $B[k]$ and the complex coefficients $F[k]$ is obtained as

$$
\begin{aligned}
F[k] &= \frac{1}{2}(A[k] - jB[k]) \quad k \neq 0 \\
F[0] &= A[0].
\end{aligned}
\qquad\qquad (4.143)
$$

Fig. 4.34 Graph of $f(t)$ for
Example 4.32

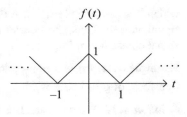

Example 4.32 Find the Fourier series representation of the periodic signal shown in Fig. 4.34.

Solution 4.32 It is seen from Fig. 4.34 that the period of the signal is 2. The signal in Fig. 4.34 can be expressed in terms of ramp signals as

$$f(t) = r(t+1) - 2r(t) + r(t-1)$$

from which the first and second derivative signals can be calculated as

$$\dot{f}(t) = u(t+1) - 2u(t) + u(t-1)$$
$$\ddot{f}(t) = \delta(t+1) - 2\delta(t) + \delta(t-1)$$

The Fourier series coefficients of the second derivative signal can be easily calculated, since it only contains impulses, and using (4.140) we can find the Fourier series coefficients of $f(t)$. We leave the rest of the solution to the reader.

Example 4.33 Derive a mathematical expression for the Fourier transform of a periodic signal $f(t)$.

Solution 4.33 Fourier series representation of $f(t)$ can be written as

$$f(t) = \sum_{k=-\infty}^{\infty} F[k] e^{jkw_0 t} \qquad w_0 = \frac{2\pi}{T} \qquad (4.144)$$

where using the Fourier transform pair

$$e^{jw_0 t} \overset{FT}{\leftrightarrow} \sqrt{2\pi}\delta(w - w_0) \qquad (4.145)$$

and evaluating the Fourier transform of both sides of (4.144), we obtain

$$\widehat{f}(w) = \sqrt{2\pi} \sum_{k=-\infty}^{\infty} F[k]\delta(w - kw_0) \qquad (4.146)$$

which is a periodic signal since Fourier series coefficients $F[k]$ is a periodic digital signal as well. Thus, the Fourier transform of a periodic signal is another periodic signal consisting of impulses.

Example 4.34 $f(t)$ is a non-periodic signal and $g(t) = f(t - t_0)$. Find the Fourier transform of $g(t)$, i.e., $\widehat{g}(w)$, in terms of the Fourier transform of $f(t)$, i.e., $\widehat{f}(w)$.

Solution 4.34 We can calculate the Fourier transform of $g(t)$ in terms of Fourier transform of $f(t)$ as

$$
\begin{aligned}
\widehat{g}(w) &= \frac{1}{\sqrt{2\pi}} \int_{-\infty}^{\infty} g(t)e^{-jwt}dt \\
&= \frac{1}{\sqrt{2\pi}} \int_{-\infty}^{\infty} f(t - t_0)e^{-jwt}dt \quad t' = t - t_0 \\
&= \frac{1}{\sqrt{2\pi}} \int_{-\infty}^{\infty} f(t')e^{-jw(t'+t_0)}dt' \\
&= e^{-jwt_0}\underbrace{\frac{1}{\sqrt{2\pi}} \int_{-\infty}^{\infty} f(t')e^{-jwt'}dt'}_{\widehat{f}(w)} \\
&= e^{-jwt_0}\widehat{f}(w).
\end{aligned}
$$

Hence, using the obtained result, we can write the transform pair

$$
f(t - t_0) \overset{\text{FT}}{\leftrightarrow} e^{-jwt_0}\widehat{f}(w) \tag{4.147}
$$

Example 4.35 Prove that

$$
tf(t) \overset{\text{FT}}{\leftrightarrow} j\frac{d\widehat{f}(w)}{dw} \tag{4.148}
$$

Solution 4.35 The Fourier transform of $f(t)$ is calculated using

$$
\widehat{f}(w) = \frac{1}{\sqrt{2\pi}} \int_{-\infty}^{\infty} f(t)e^{-jwt}dt
$$

where taking the derivative of both sides and multiplying by j, we get

$$j\frac{\widehat{df}(w)}{dw} = \frac{1}{\sqrt{2\pi}} \int\limits_{-\infty}^{\infty} tf(t)e^{-jwt}dt$$

from which obtain

$$tf(t) \overset{\text{FT}}{\leftrightarrow} j\frac{\widehat{df}(w)}{dw} \tag{4.149}$$

Example 4.36 Find the Fourier transform of $f(t) = e^{-at^2}$, $a > 0$.

Solution 4.36 To solve this problem, we will use some of the results from the previous examples. If we use the Fourier transform formula for the given signal, we obtain

$$\widehat{f}(w) = \frac{1}{\sqrt{2\pi}} \int\limits_{-\infty}^{\infty} f(t)e^{-jwt}dt \rightarrow \widehat{f}(w) = \frac{1}{\sqrt{2\pi}} \int\limits_{-\infty}^{\infty} e^{-at^2}e^{-jwt}dt$$

which cannot be further simplified. Thus, we cannot solve this question by applying the formula directly.

The first derivative of

$$f(t) = e^{-at^2}$$

can be calculated as

$$\dot{f}(t) = -2ate^{-at^2}$$

where substituting e^{-at^2} for f(t), we obtain

$$\dot{f}(t) = -2atf(t) \tag{4.150}$$

Taking the Fourier transform of both sides of (4.150), we get

$$jw\widehat{f}(w) = -2aj\frac{\widehat{df}(w)}{dw} \rightarrow \frac{\widehat{df}(w)}{\widehat{f}(w)} = -\frac{wdw}{2a}$$

where evaluating the integral of both sides, we obtain

$$\int \frac{d\widehat{f}(w)}{\widehat{f}(w)} = -\int \frac{wdw}{2a} \rightarrow \ln \widehat{f}(w) = -\frac{w^2}{4a} + c$$

from which we obtain

$$\widehat{f}(w) = Ke^{-\frac{w^2}{4a}} \tag{4.151}$$

Using (4.151), we obtain

$$\widehat{f}(w) = K\,e^{-\frac{w^2}{4a}} \rightarrow \widehat{f}(0) = K \tag{4.152}$$

Besides, we have

$$\widehat{f}(w) = \frac{1}{\sqrt{2\pi}} \int_{-\infty}^{\infty} e^{-at^2} e^{-jwt} dt \rightarrow \widehat{f}(0) = \frac{1}{\sqrt{2\pi}} \int_{-\infty}^{\infty} e^{-at^2} dt \tag{4.153}$$

Equating (4.152) to (4.153), we obtain

$$K = \frac{1}{\sqrt{2\pi}} \int_{-\infty}^{\infty} e^{-at^2} dt \tag{4.154}$$

where making use of the parameter change as in

$$u = \sqrt{a}t \quad du = \sqrt{a}dt \tag{4.155}$$

and using the property

$$\int_{-\infty}^{\infty} e^{-y^2} dy = \sqrt{\pi} \tag{4.156}$$

we obtain

$$K = \frac{1}{\sqrt{2\pi}} \int_{-\infty}^{\infty} e^{-u^2} \frac{1}{\sqrt{a}} du \rightarrow K = \frac{1}{\sqrt{2a}} \tag{4.157}$$

Example 4.37 Find the Fourier transform of $f(t) = \delta(t)$.

Solution 4.37 We can calculate the Fourier transform of the given signal as

$$\widehat{f}(w) = \frac{1}{\sqrt{2\pi}} \int_{-\infty}^{\infty} f(t) e^{-jwt} dt \rightarrow$$

$$\widehat{f}(w) = \frac{1}{\sqrt{2\pi}} \int_{-\infty}^{\infty} \delta(t) e^{-jwt} dt \rightarrow \widehat{f}(w) = \frac{1}{\sqrt{2\pi}}$$

The result can be expressed using the transform pair

$$\delta(t) \overset{FT}{\leftrightarrow} \frac{1}{\sqrt{2\pi}} \tag{4.158}$$

Example 4.38 Find the Fourier transform of $f(t) = \cos(t - 1)$.

Solution 4.38 Using the property

$$\cos(a + b) = \cos(a) \cos(b) - \sin(a) \sin(b)$$

we can expand $\cos(t - 1)$ as

$$\cos(t - 1) = \cos(t) \cos(1) + \sin(t) \sin(1) \tag{4.159}$$

Using (4.159), the Fourier transform of $f(t) = \cos(t - 1)$ can be calculated as

$$FT\{f(t)\} = \cos(1)FT\{\cos(t)\} + \sin(1)FT\{\sin(t)\} \rightarrow$$

leading to

$$\widehat{f}(w) = \cos(1)\frac{\sqrt{2\pi}}{2}[\delta(w - 1) + \delta(w + 1)] + \sin(1)\frac{\sqrt{2\pi}}{2}j[\delta(w - 1) - \delta(w + 1)]$$

where using the approximations $\cos(1) \approx 1$ and $\sin(1) \approx 0$, we obtain

$$\widehat{f}(w) \approx \frac{\sqrt{2\pi}}{2}[\delta(w - 1) + \delta(w + 1)]$$

Example 4.39 Find the Fourier transform of

$$f(t) = u(t)$$

where

$$u(t) = \begin{cases} 1 & t > 0 \\ \dfrac{1}{2} & t = 0 \\ 0 & t < 0. \end{cases} \tag{4.160}$$

Solution 4.39 $u(t)$ has a value at $t = 0$, and this complicates the evaluation of the Fourier transform. First, let us write $u(t)$ in terms of some signals whose Fourier transforms can be evaluated. Let us define $u^+(t)$ signal as

$$u^+(t) = \begin{cases} 1 & t \geq 0 \\ 0 & t < 0 \end{cases} \tag{4.161}$$

The signal $u(t)$ can be written in terms of $u^+(t)$ as

$$u(t) = \frac{1}{2} + \frac{1}{2}(u^+(t) - u^+(-t)) \tag{4.162}$$

The Fourier transform of $u^+(t)$ can be calculated as

$$\begin{aligned} \widehat{u}^+(w) &= \frac{1}{\sqrt{2\pi}} \int\limits_{t=0}^{\infty} 1 \times e^{-jwt} dt \\ &= \frac{1}{\sqrt{2\pi}} \left(-\frac{1}{jw} e^{-jwt} \Big|_{t=0}^{\infty} \right) \\ &= \frac{1}{\sqrt{2\pi}} \frac{1}{jw}. \end{aligned} \tag{4.163}$$

We have

$$u^+(-t) \overset{FT}{\leftrightarrow} \widehat{u}^+(-w)$$

Previously, we had the result

$$\delta(t) \overset{FT}{\leftrightarrow} \frac{1}{\sqrt{2\pi}} \tag{4.164}$$

for which using the duality property

$$f(t) \overset{FT}{\leftrightarrow} \widehat{f}(w) \qquad \widehat{f}(t) \overset{FT}{\leftrightarrow} f(-w) \tag{4.165}$$

we obtain

$$\frac{1}{2} \overset{FT}{\leftrightarrow} \frac{\sqrt{2\pi}}{2} \delta(w) \tag{4.166}$$

Using (4.163), (4.165), and (4.166), we can evaluate the Fourier transform of (4.162) as

$$\widehat{u}(w) = \frac{\sqrt{2\pi}}{2}\delta(w) + \frac{1}{\sqrt{2\pi}}\frac{1}{jw} \qquad (4.167)$$

If we use the formula

$$\widehat{f}(w) = \int\limits_{-\infty}^{\infty} f(t)e^{-jwt}\,dt$$

for the Fourier transform of $u(t)$, we get

$$\widehat{u}(w) = \pi\delta(w) + \frac{1}{jw} \qquad (4.168)$$

The Fourier transform $\widehat{f}(w)$ is a complex signal which has magnitude, $\left|\widehat{f}(w)\right|$, and phase, $\angle\widehat{f}(w)$, signals.

The magnitude signal $\left|\widehat{f}(w)\right|$ is also called magnitude spectral density signal, and the phase signal $\angle\widehat{f}(w)$ is also called phase spectral density signal.

Example 4.40 Find the Fourier transform of $f(t) = e^{-at}u(t)$ where $u(t)$ is the unit step signal and $a \geq 0$. After finding the Fourier transform, calculate the magnitude and phase spectrum signals.

Solution 4.40 The Fourier transform of $f(t) = e^{-at}u(t)$ can be calculated as

$$
\begin{aligned}
\widehat{f}(w) &= \frac{1}{\sqrt{2\pi}} \int\limits_{-\infty}^{\infty} f(t)e^{-jwt}\,dt \\
&= \frac{1}{\sqrt{2\pi}} \int\limits_{-\infty}^{\infty} e^{-at}u(t)e^{-jwt}\,dt \\
&= \frac{1}{\sqrt{2\pi}} \int\limits_{0}^{\infty} e^{-at}e^{-jwt}\,dt \\
&= \frac{1}{\sqrt{2\pi}} \int\limits_{0}^{\infty} e^{-t(a+jw)}\,dt \\
&= \frac{1}{\sqrt{2\pi}}\left(-\frac{1}{a+jw}\right)\left(e^{-t(a+jw)}\Big|_{0}^{\infty}\right) \frac{1}{\sqrt{2\pi}}\left(\frac{1}{a+jw}\right).
\end{aligned}
$$

Thus, the Fourier transform of $f(t)$ is

$$\widehat{f}(w) = \frac{1}{\sqrt{2\pi}} \left(\frac{1}{a+jw} \right)$$

from which the magnitude spectrum can be calculated as

$$\left| \widehat{f}(w) \right| = \frac{1}{\sqrt{2\pi}} \frac{1}{\sqrt{a^2 + w^2}}$$

The complex signal $\widehat{f}(w)$ can be written as

$$\widehat{f}(w) = \frac{1}{\sqrt{2\pi}} \frac{a - jw}{a^2 + w^2} \tag{4.169}$$

whose real and imaginary parts can be determined as

$$\widehat{f}(w) = A + jB \quad A = \frac{a}{\sqrt{2\pi}(a^2 + w^2)} \quad B = -\frac{w}{\sqrt{2\pi}(a^2 + w^2)}$$

Using (4.169), we can calculate the phase angle θ as in

$$\theta(w) = \arctan\left(\frac{B}{A} \right)$$
$$= \arctan\left(-\frac{w}{a} \right).$$

The graph of the spectral density, i.e., magnitude spectrum, is depicted in Fig. 4.35 where a is chosen as 3.

The graph of the phase spectrum signal is depicted in Fig. 4.36.

Example 4.41 The Fourier transform of a signal is given as

$$\widehat{f}(w) = 1 + e^{-jw} \tag{4.170}$$

$$\left| \widehat{f}(w) \right| = ? \quad \angle \widehat{f}(w) = ?$$

Solution 4.41 We can expand the exponential term e^{-jw} in the Fourier transform expression as

$$e^{-jw} = \cos(-w) + j\sin(-w) \tag{4.171}$$

Substituting (4.171) into (4.170), we obtain

$$\widehat{f}(w) = 1 + \cos(w) - j\sin(w) \tag{4.172}$$

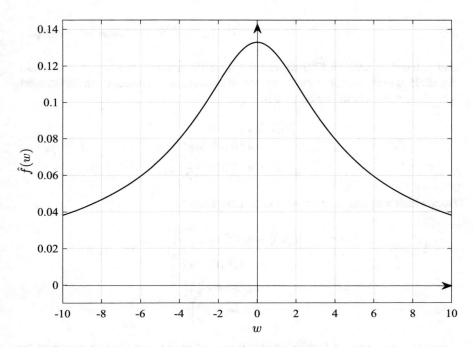

Fig. 4.35 The absolute value of the Fourier transform w.r.t. frequency variable w

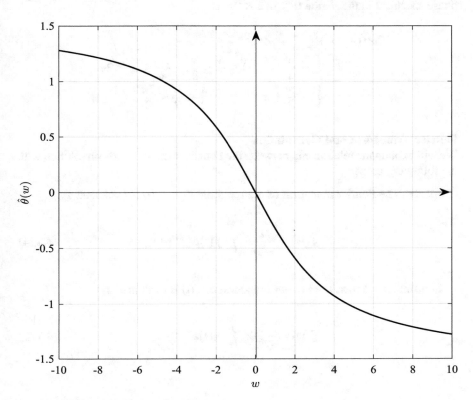

Fig. 4.36 The graph of phase spectrum signal

and proceeding with (4.172) we can calculate the magnitude and phase spectrum signal. However, to reduce the computation amount, we consider an easier solution. The Fourier transform can be written as

$$
\begin{aligned}
\widehat{f}(w) &= 1 + e^{-jw} \\
&= e^{-j\frac{w}{2}}\left(e^{j\frac{w}{2}} + e^{-j\frac{w}{2}}\right) \\
&= e^{-j\frac{w}{2}} 2 \cos\left(\frac{w}{2}\right)
\end{aligned}
\tag{4.173}
$$

from which magnitude spectrum can be calculated as

$$
\begin{aligned}
\left|\widehat{f}(w)\right| &= \left|e^{-j\frac{w}{2}} 2 \cos\left(\frac{w}{2}\right)\right| \\
&= \left|e^{-j\frac{w}{2}}\right|\left|2\cos\left(\frac{w}{2}\right)\right| \\
&= 1\left|2\cos\left(\frac{w}{2}\right)\right| \\
&= 2\left|\cos\left(\frac{w}{2}\right)\right|
\end{aligned}
$$

where we used the fact that $|e^{j\theta}| = 1$. Using (4.173), we get the one period of the phase spectrum in the region $0 \le w/2 < 2\pi$ as

$$
\begin{aligned}
\angle \widehat{f}(w) &= \angle e^{-j\frac{w}{2}} + \angle \cos\left(\frac{w}{2}\right) \\
&= \begin{cases} -\dfrac{w}{2} & 0 \le \dfrac{w}{2} \le \dfrac{\pi}{2}, \quad \dfrac{3\pi}{2} \le \dfrac{w}{2} < 2\pi \\ -\dfrac{w}{2} + \pi & \dfrac{\pi}{2} < \dfrac{w}{2} \le \dfrac{3\pi}{2} \end{cases}
\end{aligned}
$$

Fourier Transform and Convolution

We will explain the relationship between the Fourier transform and convolution with the following property.

Property The Fourier transform of non-periodic signal $f(t)$ is evaluated as

$$
\widehat{f}(w) = \frac{1}{\sqrt{2\pi}} \int_{-\infty}^{\infty} f(t)e^{-jwt}dt.
\tag{4.174}
$$

In a similar manner, the Fourier transform of $g(t)$ is evaluated as

$$
\widehat{g}(w) = \frac{1}{\sqrt{2\pi}} \int_{-\infty}^{\infty} g(t)e^{-jwt}dt.
\tag{4.175}
$$

For the Fourier transform of the convolution of these two signals, we have the transform pair

$$h(t) = f(t) * g(t) \overset{FT}{\leftrightarrow} \widehat{h}(w) = \sqrt{2\pi}\widehat{f}(w)\widehat{g}(w) \tag{4.176}$$

Proof The convolution of two signals is calculated as

$$h(t) = f(t) * g(t)$$

$$= \int_{\tau = -\infty}^{\infty} f(\tau)g(t - \tau)d\tau$$

where taking the Fourier transforms of both sides, we obtain

$$\widehat{h}(w) = \frac{1}{\sqrt{2\pi}} \int_{t = -\infty}^{\infty} \int_{\tau = -\infty}^{\infty} f(\tau)g(t - \tau)e^{-jwt}d\tau\,dt$$

where making use of the parameter change as

$$u = t - \tau \rightarrow t = u + \tau$$
$$du = dt$$

we get

$$\widehat{h}(w) = \frac{1}{\sqrt{2\pi}} \int_{\tau = -\infty}^{\infty} \int_{u = -\infty}^{\infty} f(\tau)g(u)e^{-jwu}e^{-jw\tau}du\,d\tau$$

$$= \frac{1}{\sqrt{2\pi}} \int_{\tau = -\infty}^{\infty} f(\tau)e^{-jw\tau}d\tau \int_{u = -\infty}^{\infty} g(u)e^{-jwu}du$$

$$= \widehat{f}(w)\sqrt{2\pi}\widehat{g}(w)$$

$$= \sqrt{2\pi}\,\widehat{f}(w)\widehat{g}(w).$$

Thus, we got

$$f(t) * g(t) \overset{FT}{\leftrightarrow} \sqrt{2\pi}\widehat{f}(w)\widehat{g}(w) \tag{4.177}$$

The inverse Fourier transform of

$$\sqrt{2\pi}\widehat{f}(w)\widehat{g}(w)$$

equals

$$f(t) * g(t)$$

That is,

$$f(t) * g(t) = \frac{1}{\sqrt{2\pi}} \int\limits_{w=-\infty}^{\infty} \sqrt{2\pi}\widehat{f}(w)\widehat{g}(w)e^{jwt}\,dw$$

$$= \int\limits_{w=-\infty}^{\infty} \widehat{f}(w)\widehat{g}(w)e^{jwt}\,dw.$$

Hence, we have

$$f(t) * g(t) = \int\limits_{w=-\infty}^{\infty} \widehat{f}(w)\widehat{g}(w)e^{jwt}\,dw \qquad (4.178)$$

Note The convolution of two signals can be indicated either as

$$f(t) * g(t)$$

or as

$$(f * g)(t)$$

Example 4.42 Find the Fourier transforms of $f(t) = \delta(t-1)$ and $g(t) = \delta(t-2)$. Let $h(t) = f(t) * g(t)$, and find the Fourier transform of $h(t)$.

Solution 4.42 We can calculate $\widehat{f}(w)$ and $\widehat{g}(w)$ as in

$$\widehat{f}(w) = \frac{1}{\sqrt{2\pi}} \int\limits_{-\infty}^{\infty} f(t)e^{-jwt}\,dt \qquad\qquad \widehat{g}(w) = \frac{1}{\sqrt{2\pi}} \int\limits_{-\infty}^{\infty} g(t)e^{-jwt}\,dt$$

$$= \frac{1}{\sqrt{2\pi}} \int\limits_{-\infty}^{\infty} \delta(t-1)e^{-jwt}\,dt \qquad\qquad = \frac{1}{\sqrt{2\pi}} \int\limits_{-\infty}^{\infty} \delta(t-2)e^{-jwt}\,dt$$

$$= \frac{1}{\sqrt{2\pi}}e^{-jw} \qquad\qquad\qquad\qquad = \frac{1}{\sqrt{2\pi}}e^{-jw2}$$

If $h(t) = f(t) * g(t)$, we can first calculate the convolution and then calculate the Fourier transform $\widehat{h}(w)$ as

$$
\begin{aligned}
h(t) &= \delta(t-1) * \delta(t-2) \\
&= \delta(t-3)
\end{aligned}
\qquad
\begin{aligned}
\widehat{h}(w) &= \frac{1}{\sqrt{2\pi}} \int_{-\infty}^{\infty} h(t) e^{-jwt} dt \\
&= \frac{1}{\sqrt{2\pi}} \int_{-\infty}^{\infty} \delta(t-3) e^{-jwt} dt \\
&= \frac{1}{\sqrt{2\pi}} e^{-jw3}
\end{aligned}
$$

The equality $\widehat{h}(w) = \sqrt{2\pi}\widehat{f}(w)\widehat{g}(w)$ can be verified as

$$
\frac{1}{\sqrt{2\pi}} e^{-jw3} = \sqrt{2\pi} \frac{1}{\sqrt{2\pi}} e^{-jw} \frac{1}{\sqrt{2\pi}} e^{-jw2}
$$

Example 4.43 Let $f(t) = \frac{1}{2}[\delta(t-1) + \delta(t+1)]$, and $\widehat{f}(w)$.

Solution 4.43 The Fourier transform of $f(t)$ can be calculated as

$$
\begin{aligned}
\widehat{f}(w) &= \frac{1}{\sqrt{2\pi}} \int_{-\infty}^{\infty} f(t) e^{-jwt} dt \\
&= \frac{1}{\sqrt{2\pi}} \frac{1}{2} \int_{-\infty}^{\infty} (\delta(t-1) + \delta(t+1)) e^{-jwt} dt \\
&= \frac{1}{\sqrt{2\pi}} \frac{1}{2} \left[e^{-jw} + e^{jw} \right] \\
&= \frac{1}{\sqrt{2\pi}} \cos(w).
\end{aligned}
$$

Example 4.44 The Fourier transform of a signal is given as

$$
\widehat{f}(w) = \sqrt{\pi} e^{-\frac{w^2}{4}} \cos(w) \tag{4.179}
$$

Find the time domain expression of the signal. Use the property

$$
\text{FT}\left\{ e^{-at^2} \right\} = \frac{1}{\sqrt{2a}} e^{-\frac{w^2}{4a}}
$$

for your solution.

Solution 4.44 $f(t)$ can be calculated using the inverse Fourier transform as

$$
\begin{aligned}
f(t) &= \mathrm{FT}^{-1}\left\{ \sqrt{\pi}e^{-\frac{w^2}{4}} \cos(w) \right\} \\
&= \sqrt{\pi}\frac{1}{\sqrt{2\pi}}\mathrm{FT}^{-1}\left\{ e^{-\frac{w^2}{4}} \right\} * \mathrm{FT}^{-1}\{\cos(w)\} \\
&= \sqrt{\pi}\frac{1}{\sqrt{2\pi}}\left(\sqrt{2}e^{-t^2} \right) * \left(\sqrt{2\pi}\frac{1}{2}(\delta(t-1) + \delta(t+1)) \right) \\
&= \sqrt{\frac{\pi}{2}}\left(e^{-t^2} * (\delta(t-1) + \delta(t+1)) \right) \\
&= \sqrt{\frac{\pi}{2}}\left(e^{-(t-1)^2} + e^{-(t+1)^2} \right)
\end{aligned}
$$

where we used the fact

$$
f(t) * g(t) \overset{\mathrm{FT}}{\leftrightarrow} \sqrt{2\pi}\widehat{f}(w)\widehat{g}(w)
$$

while calculating

$$
\mathrm{FT}^{-1}\left\{ e^{-\frac{w^2}{4}} \cos(w) \right\}
$$

For $\mathrm{FT}^{-1}\{\cos(w)\}$, we used the result of the previous example.

Comment Consider the transform pair

$$
\delta(t) \overset{\mathrm{FT}}{\leftrightarrow} \frac{1}{\sqrt{2\pi}}
$$

from which using the inverse Fourier transform, we can write that

$$
f(t) = \frac{1}{\sqrt{2\pi}} \int\limits_{w=-\infty}^{\infty} \widehat{f}(w)e^{jwt}\,dw \rightarrow \delta(t) = \frac{1}{\sqrt{2\pi}} \int\limits_{w=-\infty}^{\infty} \frac{1}{\sqrt{2\pi}}e^{jwt}\,dw
$$

leading to

$$
\int\limits_{w=-\infty}^{\infty} e^{jwt}\,dw = 2\pi\delta(t) \tag{4.180}
$$

which is an important equation used in the proofs of theorems. The expression in (4.180) can be stated with a different notation as

$$\int_{w=-\infty}^{\infty} e^{jw(m-n)}dw = 2\pi\delta_{m,n} \quad \delta_{m,n} = \delta(m-n). \tag{4.181}$$

Example 4.45 Show that

$$\int_{t=-\infty}^{\infty} f^2(t)dt = \int_{w=-\infty}^{\infty} \widehat{f}^2(w)dw \tag{4.182}$$

where $\widehat{f}(w)$ is the Fourier transform of $f(t)$.

Solution 4.45 $f^2(t)$ can be calculated using $f(t)f^*(t)$ where $f^*(t)$ is the conjugate of $f(t)$. Substituting

$$f(t) = \frac{1}{\sqrt{2\pi}} \int_{w=-\infty}^{\infty} \widehat{f}(w)e^{jwt}dw$$

in $f(t)f^*(t)$, we get

$$f^2(t) = f(t)f^*(t)$$

$$= \frac{1}{\sqrt{2\pi}} \int_{w=-\infty}^{\infty} \widehat{f}(w)e^{jwt}dw \left(\frac{1}{\sqrt{2\pi}} \int_{w=-\infty}^{\infty} \widehat{f}(w)e^{jwt}dw \right)^*$$

$$= \frac{1}{\sqrt{2\pi}} \int_{w=-\infty}^{\infty} \widehat{f}(w)e^{jwt}dw \left(\frac{1}{\sqrt{2\pi}} \int_{u=-\infty}^{\infty} \widehat{f}(u)e^{jut}du \right)^*$$

$$= \frac{1}{2\pi} \int_{w=-\infty}^{\infty} \int_{u=-\infty}^{\infty} \widehat{f}(w)\widehat{f}^*(u)e^{jt(w-u)}dwdu$$

where evaluating the integral of both sides w.r.t. the variable t, we obtain

$$\int_{t=-\infty}^{\infty} f^2(t)dt = \frac{1}{2\pi} \int_{t=-\infty}^{\infty} \int_{w=-\infty}^{\infty} \int_{u=-\infty}^{\infty} \widehat{f}(w)\widehat{f}^*(u)e^{jt(w-u)}du\,dw\,dt$$

where using

$$\int\limits_{t=-\infty}^{\infty} e^{jt(w-u)}dt = 2\pi\delta_{w,u} \qquad (4.183)$$

we get

$$\int\limits_{t=-\infty}^{\infty} f^2(t)dt = \frac{1}{2\pi} \int\limits_{w=-\infty}^{\infty} \int\limits_{u=-\infty}^{\infty} \widehat{f}(w)\widehat{f}^*(u)2\pi\delta_{w,u}\,du\,dw \qquad (4.184)$$

where the right side has a non-zero value for only $w = u$ and the integral leads to

$$\int\limits_{t=-\infty}^{\infty} f^2(t)dt = \frac{1}{2\pi}2\pi \int\limits_{w=-\infty}^{\infty} \widehat{f}^2(w)dw$$

which is simplified as

$$\int\limits_{t=-\infty}^{\infty} f^2(t)dt = \int\limits_{w=-\infty}^{\infty} \widehat{f}^2(w)dw \qquad (4.185)$$

which is called Parseval's relation which states that the energy of a signal can be calculated in different domains (time or frequency).

Example 4.46 We have $f(t) \overset{FT}{\leftrightarrow} \widehat{f}(w)$ and find the Fourier transform of $f\left(\frac{t}{a}\right)$.

Solution 4.46 Let $g(t) = f\left(\frac{t}{a}\right)$, and the Fourier transform of $g(t)$ can be calculated as

$$\widehat{g}(w) = \frac{1}{\sqrt{2\pi}} \int\limits_{t=-\infty}^{\infty} g(t)e^{-jwt}dt$$

$$= \frac{1}{\sqrt{2\pi}} \int\limits_{t=-\infty}^{\infty} f\left(\frac{t}{a}\right)e^{-jwt}dt$$

where defining a new parameter $u = \frac{t}{a}$ and considering the sign of a, we can evaluate the integral for $a > 0$ as

$$a>0 \quad u = \frac{t}{a} \rightarrow t = au \rightarrow dt = a\,du$$

$$\widehat{g}(w) = \frac{1}{\sqrt{2\pi}} \int_{u=-\infty}^{\infty} f(u)e^{-jwau}a\,du$$

$$= a\frac{1}{\sqrt{2\pi}} \int_{u=-\infty}^{\infty} f(u)e^{-jwau}\,du$$

$$= a\widehat{f}(aw)$$

and for $a < 0$ as

$$a < 0 \quad u = \frac{t}{a} \to t = au \to dt = adu$$

$$\widehat{g}(w) = \frac{1}{\sqrt{2\pi}} \int_{u=\infty}^{-\infty} f(u)e^{-jwau}a\,du$$

$$= -a\frac{1}{\sqrt{2\pi}} \int_{u=-\infty}^{\infty} f(u)e^{-jwau}\,du$$

$$= -a\widehat{f}(aw).$$

The results we obtained for $a > 0$ and $a < 0$ can be combined into a single mathematical expression, and we can get the transform pair

$$f\left(\frac{t}{a}\right) \overset{\text{FT}}{\leftrightarrow} |a|\widehat{f}(aw) \tag{4.186}$$

Exercise Show that

$$\text{FT}\{f^n(t)\} = (jw)^n \widehat{f}(w) \tag{4.187}$$

Property

$$\int_T e^{\pm j(k-m)\frac{2\pi}{T}t}\,dt = \begin{cases} T & k=m \\ 0 & \text{otherwise} \end{cases} \tag{4.188}$$

Example 4.47 Let $f(t)$ and $g(t)$ be periodic signals with common period T. $f(t)$ and $g(t)$ have Fourier series representations

$$f(t) = \sum_{k=-\infty}^{\infty} F[k]e^{jk\frac{2\pi}{T}t} \quad g(t) = \sum_{k=-\infty}^{\infty} G[k]e^{jk\frac{2\pi}{T}t} \tag{4.189}$$

where $F[k]$ and $G[k]$ are Fourier series coefficients. The periodic convolution of these signals is evaluated using

$$z(t) = f(t) * g(t) \rightarrow z(t) = \int_T f(\tau)g(t - \tau)d\tau \tag{4.190}$$

where $z(t)$ is a periodic signal with period T.

If we denote the Fourier series coefficients of $z(t)$ by $Z[k]$, then the relationship between $Z[k]$, $F[k]$, and $G[k]$ can be written as

$$Z[k] = TF[k]G[k] \tag{4.191}$$

Solution 4.47 Using the Fourier series representations of $f(t)$ and $g(t)$

$$f(t) = \sum_{k=-\infty}^{\infty} F[k]e^{jk\frac{2\pi}{T}t} \quad g(t) = \sum_{m=-\infty}^{m} G[m]e^{jm\frac{2\pi}{T}t}$$

we can write mathematical expressions for $f(\tau)$ and $g(t - \tau)$ in

$$z(t) = \int_T f(\tau)g(t - \tau)d\tau$$

as

$$f(\tau) = \sum_{k=-\infty}^{\infty} F[k]e^{jk\frac{2\pi}{T}\tau} \quad g(t - \tau) = \sum_{m=-\infty}^{\infty} G[m]e^{jm\frac{2\pi}{T}(t-\tau)} . \tag{4.192}$$

Substituting (4.192) into

$$z(t) = \int_T f(\tau)g(t - \tau)d\tau \tag{4.193}$$

we obtain

$$z(t) = \int_T \underbrace{\sum_{k=-\infty}^{\infty} F[k]e^{jk\frac{2\pi}{T}\tau}}_{f(\tau)} \underbrace{\sum_{m=-\infty}^{\infty} G[m]e^{jm\frac{2\pi}{T}(t-\tau)}}_{g(t-\tau)} d\tau$$

which can be arranged as

$$z(t) = \sum_{k=-\infty}^{\infty} \sum_{m=-\infty}^{\infty} F[k]G[m]e^{jm\frac{2\pi}{T}t} \int_T e^{j(k-m)\frac{2\pi}{T}\tau} d\tau$$

where using the property

$$\int_T e^{j(k-m)\frac{2\pi}{T}t} dt = \begin{cases} T & \text{if } k=m \\ 0 & \text{else} \end{cases}$$

we get

$$z(t) = \sum_{k=-\infty}^{\infty} TF[k]G[k]e^{jk\frac{2\pi}{T}t}. \tag{4.194}$$

When (4.194) is compared with the Fourier series representation of $z(t)$

$$z(t) = \sum_{k=-\infty}^{\infty} Z[k]e^{jk\frac{2\pi}{T}t}$$

we get

$$Z[k] = TF[k]G[k]$$

Thus, we obtained the transform pair

$$x(t) * y(t) \overset{\text{FSC}}{\frown\!\!\!\!\frown\!\!\!\rightarrow} TF[k]G[k] \tag{4.195}$$

where FSC means Fourier series coefficients.

Note For the previous example, if Fourier series representation of $f(t)$ and $g(t)$ are defined as

$$f(t) = \frac{1}{\sqrt{T}} \sum_{k=-\infty}^{\infty} F[k]e^{jk\frac{2\pi}{T}t} \quad g(t) = \frac{1}{\sqrt{T}} \sum_{m=-\infty}^{\infty} G[m]e^{jm\frac{2\pi}{T}t} \tag{4.196}$$

then the Fourier series coefficients of

$$z(t) = \int_T f(\tau)g(t-\tau) d\tau \tag{4.197}$$

can be calculated as

$$Z[k] = F[k]G[k] \tag{4.198}$$

Example 4.48 Let $f(t)$ and $g(t)$ be periodic signals with common period T. $f(t)$ and $g(t)$ have Fourier series representations

$$f(t) = \sum_{k=-\infty}^{\infty} F[k]e^{jk\frac{2\pi}{T}t} \quad g(t) = \sum_{k=-\infty}^{\infty} G[k]e^{jk\frac{2\pi}{T}t} \tag{4.199}$$

where $F[k]$ and $G[k]$ are Fourier series coefficients. The product of $f(t)$ and $g(t)$

$$z(t) = f(t)g(t) \tag{4.200}$$

is a periodic signal whose Fourier series coefficients are denoted by $Z[k]$. Show that

$$Z[k] = F[k] * G[k] \tag{4.201}$$

Solution 4.48 $f(t)$ and $g(t)$ have Fourier series representations

$$f(t) = \sum_{m=-\infty}^{\infty} F[m]e^{jm\frac{2\pi}{T}t} \quad g(t) = \sum_{k=-\infty}^{\infty} G[k]e^{jk\frac{2\pi}{T}t} \ .$$

The Fourier series coefficients of

$$z(t) = f(t)g(t) \tag{4.202}$$

can be calculated as

$$Z[k] = \frac{1}{T} \int_T z(t)e^{-jk\frac{2\pi}{T}t}dt \to Z[k] = \frac{1}{T} \int_T f(t)g(t)e^{-jk\frac{2\pi}{T}t}dt. \tag{4.203}$$

Substituting the Fourier series representation of $f(t)$ in (4.199) into

$$Z[k] = \frac{1}{T} \int_T f(t)g(t)e^{-jk\frac{2\pi}{T}t}dt \tag{4.204}$$

we obtain

$$Z[k] = \frac{1}{T} \int_T \sum_{m=-\infty}^{\infty} F[m]e^{jm\frac{2\pi}{T}t}g(t)e^{-jk\frac{2\pi}{T}t}dt \tag{4.205}$$

which can be further manipulated as

$$Z[k] = \frac{1}{T} \int_T \sum_{m=-\infty}^{\infty} F[m] e^{jm\frac{2\pi}{T}t} \, g(t) e^{-jk\frac{2\pi}{T}t} \, dt$$

$$= \sum_{m=-\infty}^{\infty} F[m] \underbrace{\frac{1}{T} \int_T g(t) e^{-j(k-m)\frac{2\pi}{T}t} \, dt}_{G\,[k-m]}$$

$$= \sum_{m=-\infty}^{\infty} F[m] G[k-m].$$

Thus, we obtained that

$$Z[k] = \sum_{m=-\infty}^{\infty} F[m] G[k-m] \qquad (4.206)$$

which can be written as

$$Z[k] = F[k] * G[k] \qquad (4.207)$$

Example 4.49 The Fourier transform of the periodic signal can be calculated using its Fourier series representation

$$f(t) = \sum_{k=-\infty}^{\infty} F[k] e^{jk\frac{2\pi}{T}t} \qquad (4.208)$$

as

$$\widehat{f}(w) = \sum_{k=-\infty}^{\infty} 2\pi F[k] \delta\left(w - k\frac{2\pi}{T}\right). \qquad (4.209)$$

Substituting the Fourier transform expression in (4.209) into

$$f(t) = \frac{1}{2\pi} \int_{-\infty}^{\infty} \widehat{f}(w) e^{jwt} \, dw \qquad (4.210)$$

we get

$$f(t) = \frac{1}{2\pi} \int_{-\infty}^{\infty} \sum_{k=-\infty}^{\infty} 2\pi F[k]\delta\left(w - k\frac{2\pi}{T}\right)e^{jwt}dw$$

$$= \sum_{k=-\infty}^{\infty} F[k] \int_{-\infty}^{\infty} \delta\left(w - k\frac{2\pi}{T}\right)e^{jwt}dw$$

$$= \sum_{k=-\infty}^{\infty} F[k]e^{jk\frac{2\pi}{T}t}$$

which is the Fourier series representation of $f(t)$.

Example 4.50 Show that

$$g(t) = \int_{-\infty}^{t} f(t)dt$$

equals

$$g(t) = f(t) * u(t)$$

Solution 4.50 We can expand $g(t) = f(t) * u(t)$ as

$$g(t) = \int_{-\infty}^{\infty} f(\tau)u(t - \tau)d\tau$$

where using

$$u(t - \tau) = \begin{cases} 1 & \text{if } t - \tau > 0 \rightarrow \tau < t \\ 0 & \text{otherwise} \end{cases}$$

we obtain

$$g(t) = \int_{-\infty}^{t} f(\tau)d\tau.$$

Example 4.51 If the Fourier transform of $f(t)$ is calculated using

$$F(w) = \int_{-\infty}^{\infty} f(t)e^{-jwt}dt$$

then show that we have the transform pair

$$\int_{-\infty}^{t} f(\tau)d\tau \xrightarrow{\text{Fourier Transform}} \frac{1}{jw}F(w) + \pi F(0)\delta(w) \qquad (4.211)$$

Solution 4.51 It is shown in the previous example that

$$f(t) * u(t) = \int_{-\infty}^{t} f(t)dt$$

We have

$$f(t) * u(t) \overset{\text{FT}}{\leftrightarrow} F(w)U(w)$$

in which substituting

$$U(w) = \frac{1}{jw} + \pi\delta(w)$$

we get

$$f(t) * u(t) \overset{\text{FT}}{\leftrightarrow} F(w)\left(\frac{1}{jw} + \pi\delta(w)\right)$$

which can be simplified as

$$f(t) * u(t) \overset{\text{FT}}{\leftrightarrow} \frac{1}{jw}F(w) + \pi F(0)\delta(w)$$

Hence, we obtained the transform pair

$$\int_{-\infty}^{t} f(t)dt \overset{\text{FT}}{\leftrightarrow} \frac{1}{jw}F(w) + \pi F(0)\delta(w). \qquad (4.212)$$

Example 4.52 Using the Fourier transforms of $f(t)$ and $g(t)$

$$\widehat{f}(w) = \int_{-\infty}^{\infty} f(t)e^{-jwt}dt \quad \widehat{g}(w) = \int_{-\infty}^{\infty} g(t)e^{-jwt}dt$$

show that

$$f(t)g(t) \overset{FT}{\leftrightarrow} \frac{1}{2\pi}\widehat{f}(w) * \widehat{g}(w) \tag{4.213}$$

Solution 4.52 Let $z(t) = f(t)g(t)$; the Fourier transform of $z(t)$ can be calculated as

$$\widehat{z}(w) = \int_{-\infty}^{\infty} z(t)e^{-jwt}dt \rightarrow \widehat{z}(w) = \int_{-\infty}^{\infty} f(t)g(t)e^{-jwt}dt$$

where substituting

$$g(t) = \frac{1}{2\pi} \int_{-\infty}^{\infty} \widehat{g}(w)e^{jwt}dw$$

we get

$$\widehat{z}(w) = \int_{-\infty}^{\infty} f(t) \left[\frac{1}{2\pi} \int_{-\infty}^{\infty} \widehat{g}(\lambda)e^{j\lambda t}d\lambda \right] e^{-jwt}dt$$

where rearranging the right side, we obtain

$$\widehat{z}(w) = \frac{1}{2\pi} \int_{-\infty}^{\infty} \left[\underbrace{\int_{-\infty}^{\infty} f(t)e^{-j(w-\lambda)t}d\lambda}_{\widehat{f}(w-\lambda)} \widehat{g}(\lambda)d\lambda \right]$$

from which we get

$$\widehat{z}(w) = \frac{1}{2\pi} \int_{-\infty}^{\infty} \widehat{f}(w-\lambda)\widehat{g}(\lambda)d\lambda \rightarrow \widehat{z}(w) = \frac{1}{2\pi}\widehat{f}(w) * \widehat{g}(w).$$

Example 4.53 The property

$$f(t - t_0) \overset{FT}{\leftrightarrow} e^{-jwt_0}\widehat{f}(w) \tag{4.214}$$

is valid for both periodic and non-periodic signals. Verify this property for continuous periodic signals.

Solution 4.53 The Fourier series representation of $f(t)$ is

$$f(t) = \sum_k F[k]e^{jkw_0 t} \qquad w_0 = \frac{2\pi}{T}$$

from which we obtain Fourier series representation of $g(t) = f(t - t_0)$ as

$$f(t - t_0) = \sum_k F[k]e^{jkw_0(t - t_0)}$$

leading to

$$f(t - t_0) = e^{-jkw_0 t_0} \sum_k F[k]e^{jkw_0 t}$$

whose Fourier transform can be calculated as

$$\widehat{g}(w) = e^{-jkw_0 t_0)} 2\pi \sum_k F[k]\delta(w - kw_0)$$

which implies that

$$\widehat{g}(w) = e^{-jkw_0 t_0)}\widehat{f}(w)$$

Exercise Show that the property

$$e^{jat}f(t) \overset{FT}{\leftrightarrow} \widehat{f}(w - a) \tag{4.215}$$

is valid for both continuous periodic and non-periodic signals.

Properties of Fourier Transform
In the previous section, we proved some properties of the Fourier transform. In this section, we provide the rest of the properties of the Fourier transform in Table 4.3 where the properties can be proved using the Fourier and inverse Fourier transform formulas.

Table 4.3 Fourier transform pairs

$f(t) \overset{FT}{\leftrightarrow} \widehat{f}(w)$	$g(t) \overset{FT}{\leftrightarrow} \widehat{g}(w)$
$af(t) + bg(t) \overset{FT}{\leftrightarrow} a\widehat{f}(w) + b\widehat{g}(w)$	$f(t - t_0) \overset{FT}{\leftrightarrow} e^{-jwt_0}\widehat{f}(w)$
$e^{jw_0 t}f(t) \overset{FT}{\leftrightarrow} \widehat{f}(w - w_0)$	$f^*(t) \overset{FT}{\leftrightarrow} \widehat{f}^*(-w)$
$f(at) \overset{FT}{\leftrightarrow} \frac{1}{|a|}\widehat{f}(\frac{w}{a})$	$f(t) * g(t) \overset{FT}{\leftrightarrow} \widehat{f}(w)\widehat{g}(w)$
$tf(t) \overset{FT}{\leftrightarrow} j\frac{d\widehat{f}(w)}{dw}$	$\widehat{f}(w) = \widehat{f}^*(-w)$
$\text{Even}\{f(t)\} \overset{FT}{\leftrightarrow} \text{Re}\{\widehat{f}(w)\}$	$\text{Odd}\{f(t)\} \overset{FT}{\leftrightarrow} j\text{Im}\{\widehat{f}(w)\}$
$\int\limits_{t=-\infty}^{t} f(t)dt \overset{FT}{\leftrightarrow} \frac{\widehat{f}(w)}{jw} + \sqrt{\frac{\pi}{2}}\widehat{f}(0)\delta(w)$	$\text{Re}\{\widehat{f}(w)\} = \text{Re}\{\widehat{f}^*(-w)\}$
$\text{Im}\{\widehat{f}(w)\} = \text{Im}\{\widehat{f}^*(-w)\}$	$-\angle\widehat{f}(w) = \angle\widehat{f}(-w)$
$f(t) \overset{FT}{\leftrightarrow} F(w)$	$\int\limits_{t=-\infty}^{t} f^2(t)dt = \int\limits_{w=-\infty}^{t} \widehat{f}(w)dw$
$F(t) \overset{FT}{\leftrightarrow} f(-w)$	

Summary

Fourier Series Representation

For a periodic signal $f(t)$ with period T, i.e., $f(t) = f(t + T)$, the Fourier series representation is

$$f(t) = A[0] + \sum_{k=1}^{\infty} (A[k] \cos(kw_0 t) + B[k] \sin(kw_0 t)), w_0 = \frac{2\pi}{T} \qquad (4.216)$$

where the coefficients are calculated as

$$A[0] = \frac{1}{T} \int_T f(t)dt$$

$$A[k] = \frac{2}{T} \int_T f(t) \cos(kw_0 t)dt$$

$$B[k] = \frac{2}{T} \int_T (t) \sin(kw_0 t)dt$$

The coefficients $A[0]$, $A[k]$, $B[k]$ have real values, and for this reason, they can also be called real Fourier series coefficients. The complex Fourier series representation of $f(t)$ is

$$f(t) = \sum_{k=-\infty}^{\infty} F[k]e^{jkw_0 t} \qquad w_0 = \frac{2\pi}{T} \qquad (4.217)$$

where the complex coefficients $F[k]$ are calculated as

$$F[k] = \frac{1}{T} \int\limits_{T} f(t)e^{-jkw_0t}dt$$

Fourier Integral

Fourier series representation is used for periodic signals. For non-periodic signals, Fourier integral representation is used. For the non-periodic signal $g(t)$, the Fourier integral representation is

$$g(t) = \int\limits_{0}^{T} [A(w)\cos(wt) + B(w)\sin(wt)]dw \qquad (4.218)$$

where

$$A(w) = \frac{1}{\pi} \int\limits_{-\infty}^{\infty} g(t)\cos(wt)dt$$

$$B(w) = \frac{1}{\pi} \int\limits_{-\infty}^{\infty} g(t)\sin(wt)dt$$

Fourier Transform

Fourier transform can be derived from Fourier integral representation. Fourier transform and inverse Fourier transform formulas are given as

$$g(t) = \frac{1}{\sqrt{2\pi}} \int\limits_{w=-\infty}^{\infty} \widehat{g}(w)e^{jwt}dw$$

$$\widehat{g}(w) = \frac{1}{\sqrt{2\pi}} \int\limits_{t=-\infty}^{\infty} g(t)e^{-jwt}dt \qquad (4.219)$$

The relationship between Fourier transform $\widehat{g}(w)$ and Fourier integral parameters $A(w)$ and $B(w)$ is given as

$$\widehat{g}(w) = \sqrt{\frac{\pi}{2}}(A(w) - jB(w)) \qquad (4.220)$$

On the other hand, if we define $\widehat{g}(w)$ as

$$\widehat{g}(w) = \pi(A(w) - jB(w))$$

then the Fourier transform and the inverse Fourier transform formulas are derived as

$$\widehat{g}(w) = \int_{t=-\infty}^{\infty} g(t)e^{-jwt}\,dt$$

$$g(t) = \frac{1}{2\pi} \int_{w=-\infty}^{\infty} \widehat{g}(w)e^{jwt}\,dw.$$

In this book, we use (4.219) if it is not indicated otherwise.

Fourier Transform of Periodic Signals
A periodic signal with period T has both Fourier series representation and Fourier transform. Fourier series representation of a periodic signal is given in (4.217). Fourier transform of a periodic signal is calculated using

$$\widehat{f}(w) = \sqrt{2\pi} \sum_{k=-\infty}^{\infty} F[k]\delta(w - kw_0) \qquad w_0 = \frac{2\pi}{T} \qquad (4.221)$$

If non-periodic signal $g(t)$ equals to one period of the periodic signal $f(t)$ such that

$$f(t) = \sum_{m=-\infty}^{\infty} g(t - mT)$$

then we have

$$F[k] = \frac{\sqrt{2\pi}}{T}\,\widehat{g}(w)|_{w=kw_0} \qquad w_0 = \frac{2\pi}{T} \qquad k \in Z$$

$$A[0] = \frac{\pi}{T}A(w)|_{w=0} \quad A[k] = \frac{2\pi}{T}A(w)|_{w=kw_0} \quad B[k] = \frac{2\pi}{T}A(w)|_{w=kw_0} \quad k \in Z^{+}$$

Problems

1. For $f(t) = \delta(t+1) + 2\delta(t-1)$, find and draw the graphs of $\widehat{f}(w)$ $\left|\widehat{f}(w)\right|$, $\left|\widehat{f}(w)\right|$, and $\angle \widehat{f}(w)$. Besides, find the Fourier integral representation of $f(t)$.
2. Using $f(t)$ given in (1), we obtain

$$g(t) = \sum_{k=-\infty}^{\infty} f(t - k4)$$

Find the real and complex Fourier series coefficients of $g(t)$.

3. Find the Fourier transforms of the signals:

   (a) $f(t) = te^{-t}$
   (b) $g(t) = t^2 e^{-t^2}$
   (c) $h(t) = tu(t)$
   (d) $k(t) = \cos(\alpha t)$
   (e) $l(t) = \sin(\alpha t)$
   (f) $m(t) = e^{-t} \sin(\alpha t)$
   (g) $m(t) = t \sin(\alpha t)$
   (h) $o(t) = \frac{\sin(\alpha t)}{t}$
   (i) $p(t) = e^{-|t|}$

4. $\hat{f}(w) = e^{-jw} - e^{j2w} \rightarrow \left|\hat{f}(w)\right| = ?$ $\angle \hat{f}(w) = ?$

5. Using the Fourier series representation of periodic signal $f(t)$ with period T

$$f(t) = \sum_k F[k] e^{jk\frac{2\pi}{T}t}$$

   derive the Formula used to evaluate $F[k]$.

6. Using $f(t) = u(t + 1) - 2u(t - 1) + u(t - 3)$, we obtain the periodic signal

$$g(t) = \sum_{k=-\infty}^{\infty} f(t - k8)$$

   Find the Fourier series representation of $g(t)$.

7. The impulse response of an LTI system is given as $h(t) = e^{-2t} u(t)$. Find the system outputs for the inputs:

   (a) $f(t) = \sin(\pi t)$
   (b) $f(t) = \sum_n (-1)^n \delta(t - n)$

8. Using $f(t) = \delta(t + 1) - 2\delta(t) + \delta(t - 1)$, we obtain the periodic signal $g(t)$ as

$$g(t) = \sum_{k=-\infty}^{\infty} f(t - k8)$$

   (a) Find the Fourier transform of $f(t)$.
   (b) Find the Fourier series representation of $g(t)$.

9. For the signals given below, first calculate the Fourier integral representations of these signals, and then calculate their Fourier transforms:

   (a) $f(t) = \delta(t)$
   (b) $g(t) = \delta(t + 1) + \delta(t - 1)$
   (c) $h(t) = \delta(t + 1) - \delta(t - 1)$
   (d) $l(t) = u(t) - u(t - 1)$
   (e) $i(t) = e^{-0.5t} u(t)$

10. Find the inverse Fourier transforms of:

   (a) $\widehat{f}(w) = \frac{1}{(1+jw)^2}$

   (b) $\widehat{f}(w) = \frac{2+jw}{1+jw}$

   (c) $\widehat{f}(w) = -\frac{jw}{3+j4w+(jw)^2}$

   (d) $\widehat{f}(w) = \frac{j2w+3}{6+j7w+(jw)^2}$

   (e) $\widehat{f}(w) = e^{-2w}u(w)$

   (f) $\widehat{f}(w) = \frac{\sin^2(w)}{w^2}$

11. The graphs of magnitude and phase of $\widehat{f}(w)$ are depicted in Fig. P4.1. Find the inverse Fourier transform of $\widehat{f}(w)$.

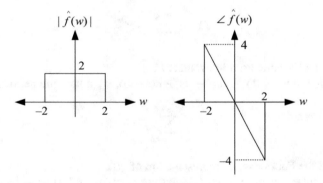

Fig. P4.1   Magnitude and phase spectrums for Problem 11

12. Find the time domain expressions of the periodic signals whose Fourier series coefficients are given as:

   (a) $F[k] = \delta(k-2) + \delta(k+2)$

   (b) $F[k] = \delta(k-1) + \delta(k+2)$

   (c) $F[k] = \left(-\frac{1}{2}\right)^{|k|}$ $w_0 = 1$

13. Calculate the Fourier transform of

$$f(t) = \sin\left(\frac{\pi}{2}t\right) + \cos\left(\frac{\pi}{3}t\right) + 1$$

Chapter 5
Fourier Analysis of Digital Signals

The advances in digital computers after the Second World War led to the appearance of a new engineering branch. Digital signal processing is the name of this engineering branch. Advances in computer technology have led to the design of almost all signal processing methods with digital technology. Analog technology has been almost completely abandoned, and most newly manufactured devices are digitally designed. Signal processing techniques developed for continuous-time signals are adapted to digital signals. Some of these digital signal processing techniques are Fourier series representation of digital signals and digital Fourier transform. The formula for the Fourier series representation of digital signals is obtained using the Fourier series representation formula of continuous-time signals. Despite advances in computer technology, some signal processing algorithms are still time-consuming and need huge computation amount for computers. For this reason, algorithms are modified to run faster by computers. One of the best known of these algorithms is the fast Fourier transform algorithm developed in the 1960s. With the development of this algorithm, there has been an acceleration in the design of systems used for signal processing.

In the world of digital technology, devices for signal processing are designed, some of which are digital signal processing chips, i.e., DSP chips and field programmable gate arrays (FPGAs). With the development of DSP chips, many algorithms implemented as hardware in the past were able to be realized as software. An example of this is the software radio. Devices produced as hardware can only be manufactured once and cannot be changed after they are produced. Devices produced using the software can be repeatedly changed and adapted to new technologies. There has been intense work on software programmable DSP and FPGA chips lately. The biggest disadvantage of software-developed devices is that they run slower than hardware-built devices. In order to alleviate this disadvantage, DSP algorithms are modified for parallel implementations. DSP and FPGA chips are used for parallel programming. Further studies focused on having DSP and FPGA chips consuming less power to enable the portability of digital devices.

© The Author(s), under exclusive license to Springer Nature Switzerland AG 2023   223
O. Gazi, *Principles of Signals and Systems*, https://doi.org/10.1007/978-3-031-17789-7_5

5.1   Fourier Series Representations of Digital Signals

5.1.1   Periodic Digital Signal

For $x[n]$, if we have $x[n] = x[n + kN]$, $N \in Z$, $k \in Z$, then $x[n]$ is a periodic signal with period N which is also called fundamental period. Every multiple of the fundamental period is another period of $x[n]$.

The angular frequency of the digital signal with period N is calculated as

$$\Omega_p = \frac{2\pi}{N} \qquad (5.1)$$

The unit of the angular frequency is the radian. In (5.1), N is an integer.

Property Let the period of $x[n]$ be N_1 and the period of $y[n]$ be N_2. The period of $h[n] = ax[n] + by[n]$ equals to the least common multiple of N_1 and N_2.

Note Sinusoidal continuous-time signals are always periodic, but the same cannot be said for digital sinusoidal signals.

Example 5.1 Is $x[n] = \sin(3n)$ a periodic signal? If it is a periodic signal, then find the period of this signal.

Solution 5.1 Using $x[n] = x[n + N]$, we obtain

$$\sin(3n) = \sin(3(n + N))$$

which can be written as

$$\sin(3n + 2\pi) = \sin(3n + 3N)$$

from which we get

$$N = \frac{2\pi}{3}$$

which is not an integer and then $x[n]$ is not a periodic signal.

5.1.2   Fourier Series Representations of Digital Signals

Let $x[n]$ be a periodic signal with period N such that $x[n] = x[n + N]$. The Fourier series representation of $x[n]$ is

$$x[n] = \frac{1}{\sqrt{N}} \sum_{k=N_0}^{N_0+N-1} X[k] e^{jk\Omega_0 n} \rightarrow x[n] = \frac{1}{\sqrt{N}} \sum_{k,N} X[k] e^{jk\Omega_0 n}, \quad \Omega_0 = \frac{2\pi}{N} \qquad (5.2)$$

where the coefficients $X[k]$ can be calculated as

$$X[k] = \frac{1}{\sqrt{N}} \sum_{n=N_0}^{N_0+N-1} x[n]e^{-jk\Omega_0 n} \rightarrow X[k] = \frac{1}{\sqrt{N}} \sum_{n,N} x[n]e^{-jk\Omega_0 n}, \quad \Omega_0 = \frac{2\pi}{N} \quad (5.3)$$

where the summation is performed over an interval containing N points. The relationship between $x[n]$ and $X[k]$ can be expressed using the transform pair

$$x[n] \xleftrightarrow{\text{DTFSC}} X[k] \qquad\qquad (5.4)$$

where DTFSC means discrete-time Fourier series coefficients.

Notation In the Fourier series representations of digital signals, we use the symbol $X[k]$ to represent the Fourier series coefficients. In some books, x_k is used instead of $X[k]$. x_k or $X[k]$ represents a complex number for each k value. In other words, x_k represents a sequence of complex numbers. The Fourier series representation formulas can also be written as

$$x[n] = \frac{1}{\sqrt{N}} \sum_{k=N_0}^{N_0+N-1} x_k e^{jk\Omega_0 n} \rightarrow x[n] = \frac{1}{\sqrt{N}} \sum_{k,N} x_k e^{jk\Omega_0 n} \quad \Omega_0 = \frac{2\pi}{N} \quad (5.5)$$

where

$$x_k = \frac{1}{\sqrt{N}} \sum_{n=N_0}^{N_0+N-1} x[n]e^{-jk\Omega_0 n} \rightarrow x_k = \frac{1}{\sqrt{N}} \sum_{n,N} x[n]e^{-jk\Omega_0 n} \quad \Omega_0 = \frac{2\pi}{N} \quad (5.6)$$

In Eqs. (5.2), (5.3), (5.5), and (5.6), summation can be evaluated from 0 to $(N-1)$. Evaluating the summation over one period for different beginnings and ends other than 0 and $(N-1)$ gives the same results.

Example 5.2 Find the Fourier series coefficients of the digital signal shown in Fig. 5.1.

Solution 5.2 The period of the signal shown in Fig. 5.1 is $N = 5$. Using the formula

Fig. 5.1 A digital signal for Example 5.2

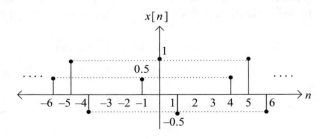

$$X[k] = \frac{1}{\sqrt{N}} \sum_{n,N} x[n] e^{-jk\Omega_0 n} \quad \Omega_0 = \frac{2\pi}{N} \tag{5.7}$$

over one-period length interval around the origin, we get

$$
\begin{aligned}
X[k] &= \frac{1}{\sqrt{5}} \sum_{n=-2}^{2} x[n] e^{-jk\Omega_0 n} \quad \Omega = \frac{2\pi}{5} \\
&= \frac{1}{\sqrt{5}} \left(x[-1] e^{-jk\frac{2\pi}{5}(-1)} + x[0] e^{-jk\frac{2\pi}{5}(0)} + x[1] e^{jk\frac{2\pi}{5}(1)} \right) \\
&= \frac{1}{\sqrt{5}} \left(0.5 e^{jk\frac{2\pi}{5}} + 1 e^0 - 0.5 e^{-jk\frac{2\pi}{5}} \right) \\
&= \frac{1}{\sqrt{5}} \left(1 + j \sin \left(k\frac{2\pi}{5} \right) \right)
\end{aligned}
$$

where we used $e^{j\theta} - e^{-j\theta} = 2j \sin(\theta)$ for the combination of the complex exponential signals. Thus, it is found that

$$X[k] = \frac{1}{\sqrt{5}} \left(1 + j \sin \left(k\frac{2\pi}{5} \right) \right)$$

which is a complex periodic signal whose period is 5. That is, the period of $X[k]$ is the same as the period of $x[n]$, i.e., we have $x[n] = x[n+5]$ and $X[k] = X[k+5]$.

Definition The complex number $X[k]$ has both magnitude and phase. The magnitude of $X[k]$, i.e., $|X[k]|$, is called the magnitude spectrum of $x[n]$. The phase of $X[k]$, $\angle X[k]$, is called the phase spectrum of $x[n]$.

Example 5.3 Obtain the Fourier series representation of

$$x[n] = \cos \left(\frac{\pi}{3} n + \phi \right) \tag{5.8}$$

Solution 5.3 Fourier series representation is available only for periodic signals; for this reason, we should first check whether the signal is periodic or not; if it is periodic, then we should find the period of the signal as a second process.

As it is stated before, every continuous-time sine signal is periodic, but the same is not true for digital sine signals. The digital cosine signal given in this question is periodic and has a period of $N = 6$. That is, we have $x[n] = x[n+6]$, as verified in

$$
\begin{aligned}
x[n] &= \cos \left(\frac{\pi}{3} n + \phi \right) \\
x[n+6] &= \cos \left(2\pi + \frac{\pi}{3} n + \phi \right) = x[n]
\end{aligned}
$$

If we use the formula directly to calculate the Fourier series coefficients, we get

$$X[k] = \frac{1}{\sqrt{6}} \sum_{n=0}^{5} \cos\left(\frac{\pi}{3}n + \phi\right) e^{-jk\Omega_0 n} \quad \Omega_0 = \frac{2\pi}{6}$$

which is not practical to calculate. We can write $x[n]$ as a sum of exponential signals as

$$
\begin{aligned}
x[n] &= \cos\left(\frac{\pi}{3}n + \phi\right) \\
&= \frac{1}{2}\left(e^{j\left(\frac{\pi}{3}n+\phi\right)} + e^{-j\left(\frac{\pi}{3}n+\phi\right)}\right) \\
&= \frac{1}{2}e^{j\phi}e^{j\frac{\pi}{3}n} + \frac{1}{2}e^{-j\phi}e^{-j\frac{\pi}{3}n} \\
&= \frac{1}{2}e^{j\phi}e^{j\frac{2\pi}{6}n} + \frac{1}{2}e^{-j\phi}e^{-j\frac{2\pi}{6}n}
\end{aligned}
$$

leading to

$$x[n] = \frac{1}{2}e^{j\phi}e^{j\frac{2\pi}{6}n} + \frac{1}{2}e^{-j\phi}e^{-j\frac{2\pi}{6}n} \tag{5.9}$$

When (5.9) is compared to the Fourier series representation of $x[n]$

$$x[n] = \frac{1}{\sqrt{6}} \sum_{k=-2}^{3} X[k]e^{jk\Omega_0 n} \quad \Omega_0 = \frac{2\pi}{6} \tag{5.10}$$

we find that

$$X[k] = \begin{cases} \dfrac{\sqrt{6}}{2}e^{j\phi} & k = 1 \\[2mm] \dfrac{\sqrt{6}}{2}e^{-j\phi} & k = -1 \\[2mm] 0 & \text{otherwise} \end{cases} \tag{5.11}$$

which can be written in terms of impulse functions as

$$X[k] = \frac{\sqrt{6}}{2}e^{j\phi}\delta[k-1] + \frac{\sqrt{6}}{2}e^{-j\phi}\delta[k+1] \tag{5.12}$$

In the previous chapter, we calculated the Fourier transforms of continuous-time sinusoidal signals, and we found that the Fourier transforms of continuous-time sinusoidal signals contain impulse functions. In (5.12), we found a similar result. However, the digital signal in (5.12) is a periodic signal, and in (5.12) its one period is depicted. The period of $X[k]$ is $N = 6$.

Exercise $x[n] = 1 + \sin\left(\frac{\pi}{12}n + \frac{3\pi}{8}\right) \rightarrow X[k] = ?$

Example 5.4 $x[n] = \sum_{l=-\infty}^{\infty} \delta[n - lN] \rightarrow X[k] = ?$

Solution 5.4 The digital signal $x[n]$ is an impulse train with period N. There is an impulse signal at the origin. If we use the formula for the calculation of the Fourier series coefficients around the origin for one-period lenh, we get

$$
\begin{aligned}
X[k] &= \frac{1}{\sqrt{N}} \sum_{n=-\frac{N-1}{2}}^{\frac{N-1}{2}} x[n] e^{-jk\frac{2\pi}{N}n} \\
&= \frac{1}{\sqrt{N}} \sum_{n=-\frac{N-1}{2}}^{\frac{N-1}{2}} \delta[n] e^{-jk\frac{2\pi}{N}n} \\
&= \frac{1}{\sqrt{N}} e^{0} \\
&= \frac{1}{\sqrt{N}}
\end{aligned}
$$

where we used the property $\sum x[n]\delta[n - n_0] = x[n_0]$ in the second line of the calculation. Using the Fourier series coefficients, we can write the Fourier series representation of $x[n]$ as

$$
\begin{aligned}
x[n] &= \frac{1}{\sqrt{N}} \sum_{k=-\frac{N-1}{2}}^{\frac{N-1}{2}} X[k] e^{jk\frac{2\pi}{N}n} \\
&= \frac{1}{\sqrt{N}} \sum_{k=-\frac{N-1}{2}}^{\frac{N-1}{2}} \frac{1}{\sqrt{N}} e^{jk\frac{2\pi}{N}n} \\
&= \frac{1}{N} \sum_{k=-\frac{N-1}{2}}^{\frac{N-1}{2}} e^{jk\frac{2\pi}{N}n}
\end{aligned}
$$

Equating the signal definition

$$
x[n] = \sum_{l=-\infty}^{\infty} \delta[n - lN]
$$

to its Fourier series representation, we obtain

$$
\sum_{l=-\infty}^{\infty} \delta[n - lN] = \frac{1}{N} \sum_{k=-\frac{N-1}{2}}^{\frac{N-1}{2}} e^{jk\frac{2\pi}{N}n} \tag{5.13}
$$

Example 5.5 Let $x[n]$ be a periodic signal with period N, and $X[k]$ are the Fourier series coefficients of $x[n]$. Find the Fourier series coefficient of

$$y[n] = e^{jk_0\frac{2\pi}{N}n}x[n]$$

in terms of $X[k]$.

Solution 5.5 We can calculate the Fourier series coefficients of $y[n]$ in terms of $X[k]$ as

$$
\begin{aligned}
Y[k] &= \frac{1}{\sqrt{N}}\sum_{n,N}y[n]e^{-jk\frac{2\pi}{N}n} \\
&= \frac{1}{\sqrt{N}}\sum_{n,N}e^{jk_0\frac{2\pi}{N}n}x[n]e^{-jk\frac{2\pi}{N}n} \\
&= \frac{1}{\sqrt{N}}\sum_{n,N}x[n]e^{-j(k-k_0)\frac{2\pi}{N}n} \\
&= X[k-k_0]
\end{aligned}
$$

Hence, we can write the transform pair

$$e^{jk_0\frac{2\pi}{N}n}x[n] \overset{\text{DTFSC}}{\longleftrightarrow} X[k-k_0] \tag{5.14}$$

where DTFSC means discrete-time Fourier series coefficients, i.e., digital Fourier series coefficients.

Example 5.6 Let $x[n]$ be a periodic signal with period N, and $X[k]$ are the Fourier series coefficients of $x[n]$. Find the Fourier series coefficients of

$$y[n] = x[n-n_0]$$

in terms of $X[k]$.

Solution 5.6 We can calculate the Fourier series coefficients of $y[n]$ in terms of $X[k]$ as

$$
\begin{aligned}
Y[k] &= \frac{1}{\sqrt{N}}\sum_{n,N}y[n]e^{-jk\frac{2\pi}{N}n} \\
&= \frac{1}{\sqrt{N}}\sum_{n,N}x[n-n_0]e^{-jk\frac{2\pi}{N}n} \quad m=n-n_0 \rightarrow n=m+n_0 \\
&= \frac{1}{\sqrt{N}}\sum_{m,N}x[m]e^{-jk\frac{2\pi}{N}(m+n_0)} \\
&= e^{-jk\frac{2\pi}{N}n_0}X[k]
\end{aligned}
$$

Thus, we obtained the transform pair

$$x[n - n_0] \overset{\text{DTFSC}}{\longleftrightarrow} e^{-jk\frac{2\pi}{N}n_0} X[k] \qquad (5.15)$$

Example 5.7 Digital signal $x[n]$ is given as

$$x[n] = \sum_{l=-\infty}^{\infty} \delta[n - l]$$

Find the Fourier series coefficients of $y[n] = x[n - 2]$.

Solution 5.7 The period of $x[n]$ is 1, and its Fourier series coefficients can be calculated as $X[k] = 1$. Using (5.15), we get

$$Y[k] = e^{-jk\frac{2\pi}{1}2} \rightarrow Y[k] = e^{-j4\pi k}$$

5.1.3 Properties for the Fourier Series Representation of Discrete-Time Periodic Signals (Digital Signals)

Let $x[n]$ and $y[n]$ be the two periodic digital signals with common period N, and let $X[k]$ and $Y[k]$ be the Fourier series coefficients of $x[n]$ and $y[n]$. We have the properties listed in Table 5.1 for these signals.

If $x[n]$ is an even signal, i.e., $x[-n] = x[n]$, then we have $X[-k] = X[k]$.

If $x[n]$ is an odd signal, i.e., $x[-n] = -x[n]$, then we have $X[-k] = -X[k]$.

5.1.4 Convolution of Periodic Digital Signals

Let $x[n]$ and $y[n]$ be the periodic signals with common period N. The periodic convolution of these two signals is calculated as

Table 5.1 Properties for DTFSCs

$x[n] \overset{\text{DTFSC}}{\longleftrightarrow} X[k], \; y[n] \overset{\text{DTFSC}}{\longleftrightarrow} Y[k]$	$ax[n] + by[n] \overset{\text{DTFSC}}{\longleftrightarrow} aX[k] + bY[k]$
$x[n - n_0] \overset{\text{DTFSC}}{\longleftrightarrow} e^{-jk\frac{2\pi}{N}n_0} X[k]$	$e^{jk\frac{2\pi}{N}n_0} x[n] \overset{\text{DTFSC}}{\longleftrightarrow} X[k - k_0]$
$x^*[n] \overset{\text{DTFSC}}{\longleftrightarrow} X^*[-k]$	$x[-n] \overset{\text{DTFSC}}{\longleftrightarrow} X[-k]$
$x[n] \otimes y[n] \overset{\text{DTFSC}}{\longleftrightarrow} X[k]Y[k]$ where $\otimes$ denotes periodic convolution	$x^*[-n] \overset{\text{DTFSC}}{\longleftrightarrow} X^*[k]$
$\sum_{m=-\infty}^{\infty} x[m] \overset{\text{DTFSC}}{\longleftrightarrow} \frac{X[k]}{1 - e^{-jk\frac{2\pi}{N}}} \quad k \neq 0$	$\sum_{n,N} |x[n]|^2 = \sum_{k,N} |X[k]|^2$

$$z[n] = x[n] \otimes y[n]$$
$$= \sum_{k=0}^{N-1} x[k]y[n-k] \tag{5.16}$$

The Fourier series coefficients of $z[n]$ can be calculated as

$$Z[k] = X[k]Y[k] \tag{5.17}$$

5.2   Fourier Transform of Digital Signals

Let $x[n]$ be a periodic signal with period N. $x[n]$ has the Fourier series representation

$$x[n] = \frac{1}{\sqrt{N}} \sum_{k,N} X[k]e^{jk\Omega_0 n} \quad \Omega_0 = \frac{2\pi}{N}$$

where

$$X[k] = \frac{1}{\sqrt{N}} \sum_{k,N} x[n]e^{-jk\Omega_0 n} \quad \Omega_0 = \frac{2\pi}{N}$$

Let one period of $x[n]$ be $x_{op}[n]$, and then we can write

$$x[n] = \sum_{l} x_{op}[n - lN] \tag{5.18}$$

which implies that

$$x_{op}[n] = \begin{cases} x[n] & 0 \le n \le N-1 \\ 0 & \text{otherwise} \end{cases} \tag{5.19}$$

Fourier series coefficients of the periodic signal in (5.18) can be calculated as

$$X[k] = \frac{1}{\sqrt{N}} \sum_{k,N} x[n]e^{-jk\Omega_0 n} \quad \Omega_0 = \frac{2\pi}{N}$$
$$= \frac{1}{\sqrt{N}} \sum_{n=-\infty}^{\infty} x_{op}[n]e^{-jk\Omega_0 n}$$

which can be further manipulated as

$$X[k] = \frac{1}{\sqrt{N}} \sum_{n=-\infty}^{\infty} x_{\mathrm{op}}[n]e^{-jk\Omega_0 n} \rightarrow \sqrt{N}X[k] = \sum_{n=-\infty}^{\infty} x_{\mathrm{op}}[n]e^{-jk\Omega_0 n}$$

where substituting $\Omega = k\Omega_0$, we obtain

$$\sqrt{N}X[k] = \sum_{n=-\infty}^{\infty} x_{\mathrm{op}}[n]e^{-jk\Omega_0 n}$$

where denoting the envelope $\sqrt{N}\, X[k]$ by $X_{\mathrm{op}}(\Omega)$, we obtain

$$X_{\mathrm{op}}(\Omega) = \sum_{n=-\infty}^{\infty} x_{\mathrm{op}}[n]e^{-j\Omega n} \tag{5.20}$$

Note that in (5.20), we have $X_{\mathrm{op}}(\Omega) = \sqrt{N}\, X[k]$ where $\Omega = k\Omega_0$. In (5.20), $X_{\mathrm{op}}(\Omega)$ is called the Fourier transform of $x_{\mathrm{op}}[n]$. We can express the result as a transform pair

$$x_{\mathrm{op}}[n] \overset{\mathrm{FT}}{\leftrightarrow} X_{\mathrm{op}}(\Omega) \tag{5.21}$$

Using

$$X[k] = \frac{1}{\sqrt{N}}X_{\mathrm{op}}(\Omega) \quad \Omega = k\Omega_0 \tag{5.22}$$

we can write the Fourier series representation of $x[n]$ as

$$\begin{aligned} x[n] &= \frac{1}{\sqrt{N}} \sum_{k,N} X[k]e^{jk\Omega_0 n} \\ &= \frac{1}{\sqrt{N}} \sum_{k,N} \frac{1}{\sqrt{N}} X_{\mathrm{op}}[\Omega]e^{jk\Omega_0 n} \end{aligned}$$

where substituting $N = \frac{2\pi}{\Omega_0}$ for N, we obtain

$$x[n] = \frac{1}{2\pi} \sum_{k,N} \Omega_0 X_{\mathrm{op}}(k\Omega_0)e^{jk\Omega_0 n} \tag{5.23}$$

As $N \rightarrow \infty$, $x[n]$ equals to $x_{op}[n]$, and the summation expression in (5.23) converges to integration, we obtain

$$x_{\mathrm{op}}[n] = \frac{1}{2\pi} \int_{2\pi} X_{\mathrm{op}}(\Omega)e^{j\Omega n} d\Omega \tag{5.24}$$

Let us write all the results we have obtained for non-periodic digital signals as

$$x_{\text{op}}[n] \overset{\text{FT}}{\leftrightarrow} X_{\text{op}}(\Omega) \qquad X_{\text{op}}(\Omega) = \sum_{n=-\infty}^{\infty} x_{\text{op}}[n]e^{-j\Omega n}$$

$$x_{\text{op}}[n] = \frac{1}{2\pi} \int_{2\pi} X_{\text{op}}(\Omega)e^{j\Omega n}d\Omega \tag{5.25}$$

Some authors denote the envelope

$$\sqrt{\frac{N}{2\pi}}X[k] \tag{5.26}$$

by $X_{\text{op}}(\Omega)$, and in this case, we have the Fourier transform and inverse Fourier transform formulas

$$x_{\text{op}}[n] \overset{\text{FT}}{\leftrightarrow} X_{\text{op}}(\Omega) \qquad X_{\text{op}}(\Omega) = \frac{1}{\sqrt{2\pi}} \sum_{n=-\infty}^{\infty} x_{\text{op}}[n]e^{-j\Omega n}$$

$$x_{\text{op}}[n] = \frac{1}{\sqrt{2\pi}} \int_{2\pi} X_{\text{op}}(\Omega)e^{j\Omega n}d\Omega \tag{5.27}$$

The Fourier transform may not exist for some non-periodic digital signals.

In order for a non-periodic signal $x_{\text{op}}[n]$ to have Fourier transform, it must satisfy one of the conditions:

1. The signal $x_{\text{op}}[n]$ must be absolutely summable, that is,

$$\sum_{n=-\infty}^{\infty} |x_{\text{op}}[n]| < \infty \tag{5.28}$$

2. The $x_{\text{op}}[n]$ must be a finite energy signal, that is,

$$\sum_{n=-\infty}^{\infty} |x_{\text{op}}[n]|^2 < \infty \tag{5.29}$$

Note There are important differences between the Fourier transform of a continuous-time signal and the Fourier transform of a digital signal. Let $x(t)$ be a continuous-time signal, and $y[n]$ be a digital signal. For these signals, the Fourier transform and inverse Fourier transform expressions are given as

$$\widehat{x}(w) = \frac{1}{\sqrt{2\pi}} \int_{w=-\infty}^{\infty} x(t)e^{-jwt}dt \quad x(t) = \frac{1}{\sqrt{2\pi}} \int_{w=-\infty}^{\infty} \widehat{x}(w)e^{jwt}dw$$

$$y(\Omega) = \frac{1}{\sqrt{2\pi}} \sum_{n=-\infty}^{\infty} y[n]e^{-j\Omega n} \qquad y[n] = \frac{1}{\sqrt{2\pi}} \int_{2\pi} y(\Omega)e^{j\Omega n}d\Omega$$

The frontiers of the integral for the Fourier transform of $x(t)$ ranges from $-\infty$ to ∞; on the other hand, the integral frontiers for the Fourier transform of $y[n]$ ranges from $k2\pi$ to $(k+1)2\pi$, for instance, from 0 to 2π. The Fourier transform of the digital signal $y[n]$, i.e., $y(\Omega)$, is a periodic signal with period 2π.

Example 5.8 Calculate the Fourier transform of

$$x[n] = \alpha^n u[n], \quad |\alpha| < 1 \tag{5.30}$$

and draw the graph of the Fourier transform.

Solution 5.8 Using the Fourier transform formula, $X(\Omega)$ can be calculated as

$$
\begin{aligned}
X(\Omega) &= \frac{1}{\sqrt{2\pi}} \sum_{n=-\infty}^{\infty} x[n]e^{-j\Omega n} \\
&= \frac{1}{\sqrt{2\pi}} \sum_{n=-\infty}^{\infty} \alpha^n u[n]e^{-j\Omega n} \\
&= \frac{1}{\sqrt{2\pi}} \sum_{n=0}^{\infty} \alpha^n e^{-j\Omega n} \\
&= \frac{1}{\sqrt{2\pi}} \sum_{n=0}^{\infty} (\alpha e^{-j\Omega})^n
\end{aligned}
$$

where employing the property

$$\sum_{n=0}^{\infty} r^n = \frac{1}{1-r}, \quad |r| < 1 \tag{5.31}$$

we obtain

$$X(\Omega) = \frac{1}{\sqrt{2\pi}} \frac{1}{1 - \alpha e^{-j\Omega}} \tag{5.32}$$

Let us draw the graphs of $|X(\Omega)|$ and $\angle X(\Omega)$ for $\alpha = \frac{1}{2}$. Substituting $\frac{1}{2}$ for α in (5.32), we get

$$X(\Omega) = \frac{1}{\sqrt{2\pi}} \frac{1}{1 - 0.5e^{-j\Omega}}$$
$$= \frac{1}{\sqrt{2\pi}} \frac{1}{1 - 0.5(\cos(\Omega) - j\sin(\Omega))}$$
$$= \frac{1}{\sqrt{2\pi}} \frac{1}{1 - 0.5\cos(\Omega) + j0.5\sin(\Omega)}$$

from which $|X(\Omega)|$ can be calculated as

$$|X(\Omega)| = \frac{1}{\sqrt{2\pi}} \frac{1}{\sqrt{(1 - 0.5\cos(\Omega))^2 + (0.5\sin(\Omega))^2}} \tag{5.33}$$

whose graph is shown in Fig. 5.2. It is also possible to draw the graph of $|X(\Omega)|$ roughly. To achieve this, we need to calculate the value of (5.33) at some critical points such as $0, \frac{\pi}{2}, \pi$ and draw the graph accordingly. $X(\Omega)$ is a periodic function, so its magnitude and phase functions are also periodic.

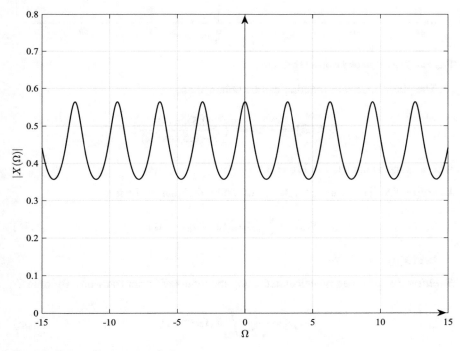

Fig. 5.2 The graph of the magnitude spectrum

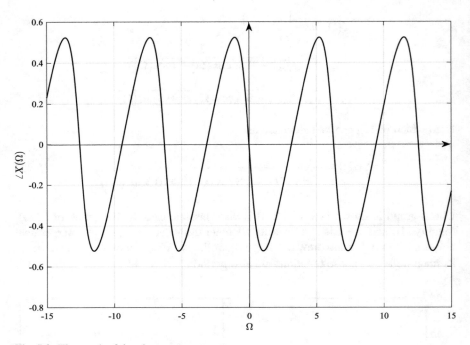

Fig. 5.3 The graph of the phase spectrum

The phase spectrum function can be obtained as

$$\angle X(\Omega) = -\arctan\left(\frac{0.5\sin(\Omega)}{1 - 0.5\cos(\Omega)}\right)$$

whose graph is depicted in Fig. 5.3.

Example 5.9 The Fourier transform of a digital signal is given as

$$X(\Omega) = \sum_k 2\pi\delta(\Omega - \Omega_0 - 2\pi k) \tag{5.34}$$

Find $x[n]$.

Solution 5.9 $x[n]$ can be calculated using the inverse Fourier transform formula

$$x[n] = \frac{1}{\sqrt{2\pi}}\int_{2\pi} X(\Omega e^{j\Omega n}d\Omega \tag{5.35}$$

In (5.35), the integral can be evaluated from 0 to 2π, i.e., over one period, and one period of $X(\Omega)$ can be calculated as

$$X(\Omega) = \sum_k 2\pi\delta(\Omega - \Omega_0 - 2\pi k) \rightarrow X(\Omega) = 2\pi\delta(\Omega - \Omega_0) \qquad (5.36)$$

When (5.36) is used in (5.35), we get

$$
\begin{aligned}
x[n] &= \frac{1}{\sqrt{2\pi}} \int_{2\pi} \sum_k 2\pi\delta(\Omega - \Omega_0 - 2\pi k)e^{j\Omega n}d\Omega \\
&= \frac{1}{\sqrt{2\pi}} \int_{2\pi} 2\pi\delta(\Omega - \Omega_0)e^{j\Omega n}d\Omega \\
&= \sqrt{2\pi}e^{j\Omega_0 n}
\end{aligned}
\qquad (5.37)
$$

The result can be expressed with a transform pair as

$$e^{j\Omega_0 n} \overset{FT}{\leftrightarrow} \sqrt{2\pi}\sum_k \delta(\Omega - \Omega_0 - 2\pi k) \qquad (5.38)$$

5.2.1   *Fourier Transform of Digital Periodic Signals*

The periodic signal $x[n]$ has the Fourier series representation

$$x[n] = \frac{1}{\sqrt{N}} \sum_{k,N} X[k]e^{jk\Omega_0 n} \qquad (5.39)$$

If we take the Fourier transforms of both sides of (5.39), we get

$$FT\{x[n]\} = \frac{1}{\sqrt{N}} \sum_{k,N} X[k]FT\{e^{jk\Omega_0 n}\} \qquad (5.40)$$

where using

$$e^{j\Omega_0 n} \overset{FT}{\leftrightarrow} \sqrt{2\pi}\sum_k \delta(\Omega - \Omega_0 - 2\pi k) \qquad (5.41)$$

we get

$$X(\Omega) = \frac{1}{\sqrt{N}} \sum_{k,N} X[k] \sum_l \sqrt{2\pi}\delta(\Omega - k\Omega_0 - 2\pi l)$$

leading to

$$X(\Omega) = \sqrt{\frac{2\pi}{N}}\sum_k X[k]\delta(\Omega - k\Omega_0) \qquad (5.42)$$

where $\Omega_0 = \frac{2\pi}{N}$.

Thus, the Fourier transform of the periodic signal $x[n]$, whose period is N, is calculated using

$$X(\Omega) = \sqrt{\frac{2\pi}{N}} \sum_k X[k]\delta(\Omega - k\Omega_0) \qquad (5.43)$$

Example 5.10 Find the Fourier transform of the periodic impulse train shown in Fig. 5.4.

Solution 5.10 The Fourier transform of the signal in Fig. 5.4 can be calculated using the transform formula

$$X(\Omega) = \sqrt{\frac{2\pi}{N}} \sum_k X[k]\delta(\Omega - k\Omega_0) \qquad (5.44)$$

To calculate (5.44), we need to first find the Fourier series coefficients $X[k]$. The Fourier series coefficients can be calculated by sampling the Fourier transform of the one period of the periodic signal as in

$$X[k] = \sqrt{\frac{2\pi}{N}} \left(X_{\mathrm{op}}(\Omega) \big|_{\Omega = k\Omega_0} \right. \qquad (5.45)$$

where $X_{\mathrm{op}}(\Omega)$ is the Fourier transform of one period of $x[n]$. One period of $x[n]$ around origin is depicted in Fig. 5.5.

Fig. 5.4 Periodic impulse train for Example 5.10

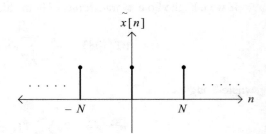

Fig. 5.5 One period of periodic impulse train

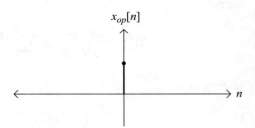

The Fourier transform of the signal in Fig. 5.5 can be calculated as

$$X_{\mathrm{op}}(\Omega) = \frac{1}{\sqrt{2\pi}} \sum_{n=-\infty}^{\infty} x_{\mathrm{op}}[n]e^{-j\Omega n} \rightarrow X_{\mathrm{op}}(\Omega) = \frac{1}{\sqrt{2\pi}} \sum_{n=-\infty}^{\infty} \delta[n]e^{-j\Omega n} \rightarrow X_{\mathrm{op}}(\Omega) = \frac{1}{\sqrt{2\pi}}$$

Fourier series coefficients can be calculated using the Fourier transform as

$$X[k] = \sqrt{\frac{2\pi}{N}} \left(X_{\mathrm{op}}(\Omega) \big|_{\Omega = k\Omega_0} \right)$$

leading to

$$X[k] = \frac{1}{\sqrt{N}} \tag{5.46}$$

Using the Fourier series coefficients in (5.46), we can calculate the Fourier transform of the periodic signal as

$$X(\Omega) = \sqrt{\frac{2\pi}{N}} \sum_{k} X[k]\delta(\Omega - k\Omega_0)$$

leading to

$$X(\Omega) = \sqrt{\frac{2\pi}{N}} \sum_{k} \frac{1}{\sqrt{N}} \delta(\Omega - k\Omega_0)$$

which can be simplified as

$$X(\Omega) = \frac{\sqrt{2\pi}}{N} \sum_{k} \delta(\Omega - k\Omega_0) \tag{5.47}$$

Thus, the Fourier transform of an impulse train with period N is another impulse train.

Example 5.11 The aperiodic digital signal $x_{\mathrm{op}}[n]$ is given as

$$x_{\mathrm{op}}[n] = \begin{cases} 1 & -M \leq n \leq M \\ 0 & \text{otherwise} \end{cases} \tag{5.48}$$

Using $x_{\mathrm{op}}[n]$, the periodic signal $x[n]$ with period N is obtained as

$$x[n] = \sum_{k} x_{\mathrm{op}}[n - kN] \quad N > 2M + 1 \tag{5.49}$$

The graphs of $x_{\mathrm{op}}[n]$ and $x[n]$ are depicted in Fig. 5.6. Calculate the Fourier series coefficients of $x[n]$.

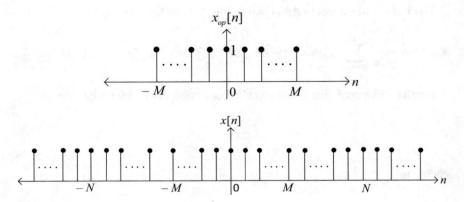

Fig. 5.6 Digital signals for Example 5.11

Solution 5.11 We can calculate the Fourier series coefficients of the periodic signal shown in Fig. 5.6 as

$$
\begin{aligned}
X[k] &= \frac{1}{\sqrt{N}} \sum_{n=-M}^{M} x[n] e^{-jk\Omega_0 n} \\
&= \frac{1}{\sqrt{N}} \sum_{n=-M}^{M} e^{-jk\Omega_0 n} \quad m = n + M \\
&= \frac{1}{\sqrt{N}} \sum_{m=0}^{2M} \left(e^{-jk\Omega_0 m} \right) e^{jk\Omega_0 M} \\
&= \frac{e^{jk\Omega_0 M}}{\sqrt{N}} \left(\frac{1 - e^{-jk\Omega_0(2M+1)}}{1 - e^{-jk\Omega_0}} \right).
\end{aligned}
$$

Thus, we obtained that

$$
X[k] = \frac{e^{jk\Omega_0 M}}{\sqrt{N}} \left(\frac{1 - e^{-jk\Omega_0(2M+1)}}{1 - e^{-jk\Omega_0}} \right) \tag{5.50}
$$

The result in (5.50) contains exponential functions. We can manipulate the mathematical expression in (5.50) and express the result in terms of the sinusoidal functions as

$$
\begin{aligned}
X[k] &= \frac{e^{jk\Omega_0 M}}{\sqrt{N}} \left(\frac{1 - e^{-jk\Omega_0(2M+1)}}{1 - e^{-jk\Omega_0}} \right) \\
&= \frac{e^{jk\Omega_0 M}}{\sqrt{N}} \frac{e^{-\frac{jk\Omega_0(2M+1)}{2}}}{e^{-\frac{jk\Omega_0}{2}}} \left(\frac{e^{\frac{jk\Omega_0(2M+1)}{2}} - e^{-\frac{jk\Omega_0(2M+1)}{2}}}{e^{\frac{jk\Omega_0}{2}} - e^{-\frac{jk\Omega_0}{2}}} \right)
\end{aligned}
$$

leading to

$$X[k] = \frac{1}{\sqrt{N}} \frac{\sin\left((k\Omega_0(2M+1)/2\right)}{\sin\left(\frac{k\Omega_0}{2}\right)} \tag{5.51}$$

Note

$$
\begin{aligned}
1 - e^{j\Omega} &= e^{\frac{j\Omega}{2}}\left(e^{-\frac{j\Omega}{2}} - e^{\frac{j\Omega}{2}}\right) \\
&= -2je^{\frac{j\Omega}{2}} \sin\left(\frac{\Omega}{2}\right) \\
1 + e^{j\Omega} &= e^{\frac{j\Omega}{2}}\left(e^{-\frac{j\Omega}{2}} + e^{\frac{j\Omega}{2}}\right) \\
&= 2e^{\frac{j\Omega}{2}} \cos\left(\frac{\Omega}{2}\right)
\end{aligned}
\tag{5.52}
$$

Substituting $\Omega_0 = \frac{2\pi}{N}$ in $X[k]$, we obtain

$$X[k] = \frac{1}{\sqrt{N}} \frac{\sin\left(\frac{k\pi(2M+1)}{N}\right)}{\sin\left(\frac{k\pi}{N}\right)} \tag{5.53}$$

where denominator equals 0 for $k = \pm N, \pm 2N, \dots$ To evaluate the value of $X[k]$ for $k = \pm N, \pm 2N, \dots$ we can apply the L'Hospital's rule as

$$\lim_{k \to 0, \pm N, \dots} \frac{1}{\sqrt{N}} \frac{\sin\left(\frac{k\pi(2M+1)}{N}\right)}{\sin\left(\frac{k\pi}{N}\right)} = \frac{2M+1}{\sqrt{N}} \tag{5.54}$$

Using the evaluated Fourier series coefficients, we can write the Fourier series representation of the periodic signal as

$$
\begin{aligned}
x[n] &= \frac{1}{\sqrt{N}} \sum_{k=-\frac{N}{2}}^{\frac{N}{2}} X[k] e^{jk\frac{2\pi}{N}n} \\
&= \frac{1}{\sqrt{N}} \sum_{k=-\frac{N}{2}}^{\frac{N}{2}} \frac{1}{\sqrt{N}} \frac{\sin\left(\frac{k\pi(2M+1)}{N}\right)}{\sin\left(\frac{k\pi}{N}\right)} e^{jk\frac{2\pi}{N}n} \\
&= \frac{1}{N} \sum_{k=-\frac{N}{2}}^{\frac{N}{2}} \frac{\sin\left(\frac{k\pi(2M+1)}{N}\right)}{\sin\left(\frac{k\pi}{N}\right)} e^{jk\frac{2\pi}{N}n}
\end{aligned}
$$

Hence, we obtained that

$$x[n] = \frac{1}{N} \sum_{k=-\frac{N}{2}}^{\frac{N}{2}} \frac{\sin\left(\frac{k\pi(2M+1)}{N}\right)}{\sin\left(\frac{k\pi}{N}\right)} e^{jk\frac{2\pi}{N}n} \tag{5.55}$$

We can calculate the Fourier series coefficients of the periodic signal using an alternative approach. In this alternative approach, we first calculate the Fourier transform of one period of $x[n]$ using

$$X_{op}(\Omega) = \frac{1}{\sqrt{2\pi}} \sum_n x_{op}[n] e^{-j\Omega n}$$

leading to

$$X_{op}(\Omega) = \frac{1}{\sqrt{2\pi}} \frac{\sin\left(\frac{\Omega(2M+1)}{2}\right)}{\sin\left(\frac{\Omega}{2}\right)} \tag{5.56}$$

$X[k]$ can be obtained from $X_{op}(\Omega)$ by sampling operation as described in

$$X[k] = \sqrt{\frac{2\pi}{N}} X_{op}(\Omega)\big|_{\Omega = k\Omega_0} \qquad \Omega_0 = \frac{2\pi}{N}$$

The graphs of $X_{op}(\Omega)$ and $X[k]$ are depicted in Fig. 5.7 for $M = 2$ and $N = 40$. It is seen from Fig. 5.7 that the Fourier series coefficients are the scaled Fourier transform samples.

If we had defined the formula for the calculation of Fourier series coefficients as

$$X[k] = \sum_n x[n] e^{-jk\Omega_0 n} \tag{5.57}$$

then the Fourier the relationship between Fourier transform and Fourier series coefficients can be written as

$$X[k] = X_{op}(\Omega)\big|_{\Omega = k\Omega_0} \qquad \Omega_0 = \frac{2\pi}{N} \tag{5.58}$$

and in this case, the Fourier series coefficients become equal to the samples of the Fourier transform of the one period of the periodic signal as shown in Fig. 5.8. It is seen from Fig. 5.8 that $X[k]$ values are samples of $X_{op}(\Omega)$. The period of $X_{op}(\Omega)$ is 2π. The period of the digital signal $X[k]$ is $N = 40$.

Example 5.12 $X(\Omega)$ is periodic with period 2π. One period of $X(\Omega)$ is given as

$$X(\Omega) = \begin{cases} 1 & |\Omega| \leq W \\ 0 & W < |\Omega| < \pi. \end{cases} \tag{5.59}$$

Determine $x[n]$ whose Fourier transform is $X(\Omega)$.

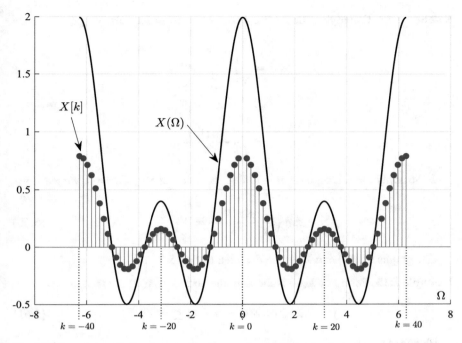

Fig. 5.7 The relationship between the Fourier transform and the Fourier series coefficients

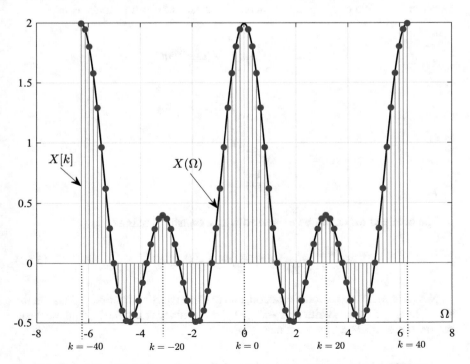

Fig. 5.8 The relationship between the Fourier transform and the Fourier series coefficients

Solution 5.12 Employing the inverse Fourier transform formula, we get

$$
\begin{aligned}
x[n] &= \frac{1}{\sqrt{2\pi}} \int\limits_{-W}^{W} e^{j\Omega n} d\Omega \\
&= \frac{1}{2\pi} \left(\frac{1}{jn} e^{j\Omega n} \Big|_{-W}^{W} \right) \\
&= \sqrt{\frac{2}{\pi}} \frac{\sin(Wn)}{n}.
\end{aligned}
$$

If we apply the L'Hospital's rule to find the value of $x[n]$ at $n = 0$, we get

$$
\lim_{n \to 0} \sqrt{\frac{2}{\pi}} \frac{\sin(Wn)}{n} = W \sqrt{\frac{2}{\pi}} \tag{5.60}
$$

The graphs of $x[n]$ and $X(\Omega)$ are depicted in Fig. 5.9.

Example 5.13 Period of $X(\Omega)$ is 2π, and one period of $X(\Omega)$ is given as

$$
X(\Omega) = \delta(\Omega) - \pi < \Omega \leq \pi \tag{5.61}
$$

Find $x[n]$.

Solution 5.13 We can employ the inverse Fourier transform formula as in

$$
\begin{aligned}
x[n] &= \frac{1}{\sqrt{2\pi}} \int\limits_{-\pi}^{\pi} X(\Omega) e^{j\Omega n} d\Omega \\
&= \frac{1}{\sqrt{2\pi}} \int\limits_{-\pi}^{\pi} \delta(\Omega) e^{j\Omega n} d\Omega \\
&= \frac{1}{\sqrt{2\pi}} e^{0} \\
&= \frac{1}{\sqrt{2\pi}}.
\end{aligned}
$$

The obtained result can be expressed with a transform pair as

$$
\frac{1}{\sqrt{2\pi}} \overset{FT}{\leftrightarrow} \delta(\Omega) \quad -\pi < \Omega \leq \pi \tag{5.62}
$$

Note that $X(\Omega)$ is a periodic function, and in (5.62) only one period of $X(\Omega)$ on the interval $-\pi < \Omega \leq \pi$ is written. If $-\pi < \Omega \leq \pi$ is omitted from (5.62), then the result can be expressed as a transform pair

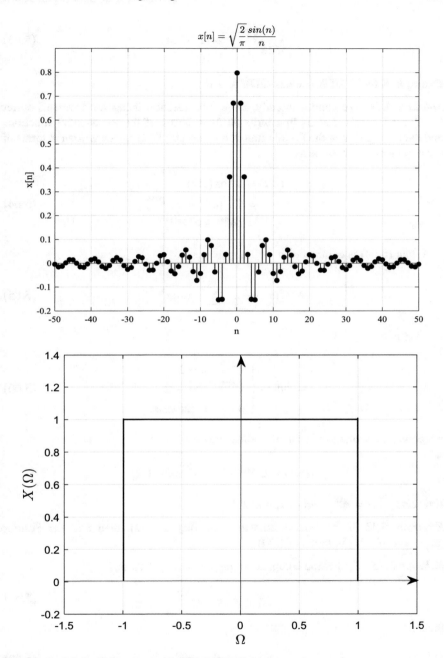

Fig. 5.9 Scaled sinc(·) function and its Fourier transform

$$\frac{1}{\sqrt{2\pi}} \overset{FT}{\leftrightarrow} \sum_k \delta(\Omega - k2\pi) \tag{5.63}$$

Example 5.14 If $X(\Omega) = 2\cos(2\Omega)$, find $x[n]$.

Solution 5.14 We cannot directly solve this question using the inverse Fourier transform formula. Let us first write $X(\Omega)$ in terms of the exponential functions and then compare it to the Fourier transform formula. $X(\Omega)$ can be written in terms of the exponential functions as

$$
\begin{aligned}
X(\Omega) &= 2\cos(2\Omega) \\
&= 2\frac{1}{2}\left(e^{j2\Omega} + e^{-j2\Omega}\right) \\
&= \left(e^{j2\Omega} + e^{-j2\Omega}\right)
\end{aligned}
\tag{5.64}
$$

When (5.64) is compared to

$$X(\Omega) = \frac{1}{\sqrt{2\pi}} \sum_{n=-\infty}^{\infty} x[n] e^{-j\Omega n} \tag{5.65}$$

we find that

$$
x[n] = \begin{cases}
\sqrt{2\pi} & n = 2 \\
\sqrt{2\pi} & n = -2 \\
0 & \text{otherwise}
\end{cases}
\tag{5.66}
$$

which can be written in terms of impulse functions as

$$x[n] = \sqrt{2\pi}\delta[n-2] + \sqrt{2\pi}\delta[n+2]$$

Exercise $x[n] = a^{|n|}$, $|a| < 1$, find $X(\Omega)$.

Example 5.15 If the Fourier transform of $x[n]$ is $X(\Omega)$, then find the Fourier transform of $x^*[n]$ in terms of $X(\Omega)$.

Solution 5.15 $x^*[n]$ is the complex conjugate of $x[n]$. That is, if

$$x[n] = a[n] + jb[n] \tag{5.67}$$

then

$$x^*[n] = a[n] - jb[n] \tag{5.68}$$

Let $y[n] = x^*[n]$; the Fourier transform of $y[n]$ can be calculated as

$$Y(\Omega) = \frac{1}{\sqrt{2\pi}} \sum_{n=-\infty}^{\infty} y[n]e^{-j\Omega n}$$

where substituting $y[n] = x^*[n]$, we obtain

$$Y(\Omega) = \frac{1}{\sqrt{2\pi}} \sum_{n=-\infty}^{\infty} x^*[n]e^{-j\Omega n} \qquad (5.69)$$

where taking the complex conjugate of both sides, we get

$$Y^*(\Omega) = \frac{1}{\sqrt{2\pi}} \sum_{n=-\infty}^{\infty} x[n]e^{j\Omega n} \qquad (5.70)$$

Note that for complex signals $w[n]$ and $z[n]$, we have the properties

$$(w[n] + z[n])^* = w^*[n] + z^*[n]$$

and

$$(w[n]z[n])^* = w^*[n]z^*[n]$$

Using (5.70), we get

$$Y^*(-\Omega) = \frac{1}{\sqrt{2\pi}} \underbrace{\sum_{n=-\infty}^{\infty} x[n]e^{-j\Omega n}}_{X(\Omega)} \qquad (5.71)$$

where the right-hand side is $X(\Omega)$. Thus we showed that

$$Y^*(-\Omega) = X(\Omega)$$

from which we obtain

$$Y(\Omega) = X^*(-\Omega) \qquad (5.72)$$

The result can be expressed with a transform pair as

$$x^*[n] \overset{FT}{\leftrightarrow} X^*(-\Omega) \qquad (5.73)$$

Example 5.16 Verify the equality

$$\sum_{n=0}^{N-1} e^{jk\frac{2\pi}{N}n} = \begin{cases} N & \text{if } k = \pm mN \\ 0 & \text{otherwise} \end{cases} \tag{5.74}$$

Solution 5.16 In calculus, we have the property

$$\sum_{n=0}^{N-1} x^n = \frac{1-x^N}{1-x} \tag{5.75}$$

where substituting $x = e^{jk\frac{2\pi}{N}}$ for the left side, we get

$$\sum_{n=0}^{N-1} e^{jk\frac{2\pi}{N}n}$$

which can be evaluated for $k \neq mN$ as

$$\sum_{n=0}^{N-1} e^{jk\frac{2\pi}{N}n} = \frac{1 - \left(e^{jk\frac{2\pi}{N}}\right)^N}{1 - e^{jk\frac{2\pi}{N}}} \rightarrow = \frac{1 - \overbrace{e^{jk2\pi}}^{=1}}{1 - e^{jk\frac{2\pi}{N}}} \rightarrow 0$$

and for $k = mN$ we have

$$\sum_{n=0}^{N-1} e^{jmN\frac{2\pi}{N}n} \rightarrow \sum_{n=0}^{N-1} \underbrace{e^{jm2\pi n}}_{=1} = N$$

Combining the obtained results, we get

$$\sum_{n=0}^{N-1} e^{jk\frac{2\pi}{N}n} = \begin{cases} N & \text{if } k = \pm mN \\ 0 & \text{otherwise} \end{cases} \tag{5.76}$$

which can be expressed as

$$\sum_{n=0}^{N-1} e^{jk\frac{2\pi}{N}n} = N\delta[k \bmod N] \tag{5.77}$$

Example 5.17 $x[n]$ and $y[n]$ are two periodic digital signals with period N. Let $z[n]$ be the periodic convolution of these two signals, i.e.,

$$z[n] = \sum_{m,N} x[m]y[n-m] \tag{5.78}$$

Show that the relationship between the Fourier series coefficient of $z[n]$ and the Fourier series coefficients of $x[n]$, $y[n]$ can be written as

$$Z[k] = X[k]Y[k] \tag{5.79}$$

Solution 5.17 Substituting the Fourier series representation of $y[n]$

$$y[n] = \frac{1}{\sqrt{N}} \sum_{n,N} Y[k]e^{jk\frac{2\pi}{N}n}$$

into the periodic convolution expression

$$z[n] = \sum_{m,N} x[m]y[n-m]$$

we get

$$z[n] = \sum_{m,N} x[m] \frac{1}{\sqrt{N}} \sum_{n,N} Y[k]e^{jk\frac{2\pi}{N}(n-m)}$$

which can be further manipulated as

$$z[n] = \frac{1}{\sqrt{N}} \sum_{n,N} \underbrace{\left[\sum_{m,N} x[m]e^{-jk\frac{2\pi}{N}m} \right]}_{\sqrt{N}X[k]} Y[k]e^{jk\frac{2\pi}{N}n}$$

$$= \sum_{n,N} \underbrace{X[k]Y[k]e^{jk\frac{2\pi}{N}n}}_{Z[k]}$$

from which we obtain that

$$Z[k] = X[k]Y[k] \tag{5.80}$$

The result can be written with a transform pair as

$$\sum_{m,N} x[m]y[n-m] \overset{\text{FSC}}{\leftrightarrow} X[k]Y[k] \tag{5.81}$$

Example 5.18 $x[n]$ and $y[n]$ are two periodic digital signals with period N, and $z[n]$ is the product of these two signals, i.e.,

$$z[n] = x[n]y[n] \tag{5.82}$$

Show that the relationship between the Fourier series coefficient of the $z[n]$ and the Fourier series coefficients of $x[n]$ and $y[n]$ is as

$$Z[k] = \frac{1}{\sqrt{N}} \sum_{m,N} X[m]Y[k-m] \tag{5.83}$$

That is, $Z[k]$ is equal to the scaled periodic convolution of $X[k]$ and $Y[k]$.

Solution 5.18 The Fourier series coefficients of $z[n] = x[n]y[n]$ can be calculated using

$$Z[k] = \frac{1}{\sqrt{N}} \sum_{n,N} z[n] e^{-jk\frac{2\pi}{N}n}$$

where substituting $x[n]y[n]$ for $z[n]$, we get

$$Z[k] = \frac{1}{\sqrt{N}} \sum_{n,N} x[n]y[n] e^{-jk\frac{2\pi}{N}n}$$

where substituting

$$y[n] = \frac{1}{\sqrt{N}} \sum_{m,N} Y[m] e^{jm\frac{2\pi}{N}n}$$

we obtain

$$Z[k] = \frac{1}{\sqrt{N}} \sum_{n,N} x[n] \underbrace{\left[\frac{1}{\sqrt{N}} \sum_{m,N} Y[m] e^{jm\frac{2\pi}{N}n} \right]}_{y[n]} e^{-jk\frac{2\pi}{N}n}$$

$$= \frac{1}{N} \sum_{m,N} Y[m] \underbrace{\left[\sum_{n,N} x[n] e^{-j(k-m)\frac{2\pi}{N}n} \right]}_{= \sqrt{N} X[k-m]}$$

$$= \frac{\sqrt{N}}{N} \sum_{m,N} Y[m]X[k-m]$$

$$= \frac{1}{\sqrt{N}} \sum_{m,N} X[k-m]Y[m]$$

which is nothing but the scaled periodic convolution of $X[k]$ and $Y[k]$.

Example 5.19 Show that the summation

$$\frac{1}{\sqrt{N}}\sum_{k,N}X[k]e^{jk\frac{2\pi}{N}n}$$

result equals $x[n]$.

Hint For $X[k]$, substitude

$$\frac{1}{\sqrt{N}}\sum_{m,N}x[m]e^{-jk\frac{2\pi}{N}m}$$

Solution 5.19 If

$$X[k]=\frac{1}{\sqrt{N}}\sum_{m,N}x[m]e^{-jk\frac{2\pi}{N}m}$$

is used in

$$\frac{1}{\sqrt{N}}\sum_{k,N}X[k]e^{jk\frac{2\pi}{N}n}$$

we obtain

$$\frac{1}{\sqrt{N}}\sum_{k,N}\left[\frac{1}{\sqrt{N}}\sum_{m,\,N}x[m]e^{-jk\frac{2\pi}{N}m}\right]e^{jk\frac{2\pi}{N}n}$$

which can be written as

$$\frac{1}{\sqrt{N}}\sum_{m,N}\frac{1}{\sqrt{N}}x[m]\underbrace{\sum_{k,N}e^{j(n-m)k\frac{2\pi}{N}}}_{N\delta[(n-m)\bmod N]}$$

leading to

$$\frac{1}{\sqrt{N}}\sum\frac{1}{\sqrt{N}}x[n-rN]N\delta[r]$$

which is equal to

$$x[n]$$

Thus, we showed that

$$\frac{1}{\sqrt{N}} \sum_{k,N} X[k] e^{jk\frac{2\pi}{N}n} = x[n]$$

Note If $r = (n - m) \bmod N$, then $m = n - rN$ and we have

$$\sum_{k,N} e^{j(n-m)k\frac{2\pi}{N}} = N\delta[r] \tag{5.84}$$

Example 5.20 $x[n]$ and $y[n]$ are two periodic digital signals with period N, and $z[n]$ is the product of these two signals, i.e.,

$$z[n] = x[n]y[n] \tag{5.85}$$

Show that the relationship between the Fourier transform of $z[n]$ and the Fourier transforms of $x[n]$ and $y[n]$ is as

$$Z(\Omega) = \frac{1}{\sqrt{2\pi}} \int_{2\pi} X(\lambda) Y(\Omega - \lambda) d\lambda \tag{5.86}$$

That is, $Z(\Omega)$ equals to the periodic convolution of $X(\Omega)$ and $Y(\Omega)$.

Solution 5.20 The Fourier transform of $z[n]$ can be calculated using

$$Z(\Omega) = \frac{1}{\sqrt{2\pi}} \sum_{n=-\infty}^{\infty} z[n] e^{-j\Omega n}$$

where substituting $x[n]y[n]$ for $z[n]$, we get

$$Z(\Omega) = \frac{1}{\sqrt{2\pi}} \sum_{n=-\infty}^{\infty} x[n]y[n] e^{-j\Omega n}$$

where using

$$x[n] = \frac{1}{\sqrt{2\pi}} \int_{2\pi} X(\lambda) e^{j\lambda n} d\lambda$$

we obtain

$$Z(\Omega) = \frac{1}{\sqrt{2\pi}} \sum_{n=-\infty}^{\infty} \underbrace{\left[\frac{1}{2\pi} \int_{2\pi} X(\lambda) e^{j\lambda n} d\lambda \right]}_{x[n]} y[n] e^{-j\Omega n}$$

which can be rearranged as

$$Z(\Omega) = \frac{1}{2\pi} \int_{2\pi} X(\lambda) \underbrace{\left[\sum_{n=-\infty}^{\infty} y[n] e^{-j(\Omega-\lambda)n} \right]}_{\sqrt{2\pi} Y(\Omega-\lambda)} d\lambda$$

leading to

$$Z(\Omega) = \frac{1}{\sqrt{2\pi}} \int_{2\pi} X(\lambda) Y(\Omega-\lambda) d\lambda$$

where

$$\int_{2\pi} X(\lambda) Y(\Omega-\lambda) d\lambda$$

is the periodic convolution of $X(\Omega)$ and $Y(\Omega)$. The result can be expressed with a transform pair as

$$x[n]y[n] \overset{FT}{\leftrightarrow} \frac{1}{\sqrt{2\pi}} X(\Omega) * Y(\Omega) \tag{5.87}$$

Example 5.21 Parseval's relation for digital signals is given as

$$\sum_{n=-\infty}^{\infty} |x[n]|^2 = \int_{2\pi} |X(\Omega)| d\Omega \tag{5.88}$$

Verify this equation.

Solution 5.21 The summation expression

$$\sum_{n=-\infty}^{\infty} |x[n]|^2$$

can be written as

$$\sum_{n=-\infty}^{\infty} |x[n]|^2 = \sum_{n=-\infty}^{\infty} x[n]x^*[n]$$

where substituting

$$x^*[n] = \frac{1}{\sqrt{2\pi}} \int_{2\pi} X^*(\Omega) e^{-j\Omega n} d\Omega$$

we obtain

$$\sum_{n=-\infty}^{\infty} |x[n]|^2 = \sum_{n=-\infty}^{\infty} x[n] \underbrace{\left[\frac{1}{\sqrt{2\pi}} \int_{2\pi} X^*(\Omega) e^{-j\Omega n} d\Omega \right]}_{x^*[n]}$$

where the right side can be further manipulated as

$$\sum_{n=-\infty}^{\infty} |x[n]|^2 = \frac{1}{\sqrt{2\pi}} \int_{2\pi} X^*(\Omega) \underbrace{\left[\sum_{n=-\infty}^{\infty} x[n] e^{-j\Omega n} \right]}_{\sqrt{2\pi} X(\Omega)} d\Omega$$

leading to

$$\sum_{n=-\infty}^{\infty} |x[n]|^2 = \int_{2\pi} X^*(\Omega) X(\Omega) d\Omega$$

resulting in

$$\sum_{n=-\infty}^{\infty} |x[n]|^2 = \int_{2\pi} |X(\Omega)|^2 d\Omega \tag{5.89}$$

5.2.2 Properties of the Fourier Transform

Let the Fourier transforms of $x[n]$ and $y[n]$ be $X(\Omega)$ and $Y(\Omega)$. The properties of the Fourier transform are listed in Table 5.2.

Example 5.22 If $x[n] \overset{FT}{\leftrightarrow} X(\Omega)$, then show that

$$nx[n] \overset{FT}{\leftrightarrow} j\frac{dX(\Omega)}{d\Omega} \tag{5.90}$$

Solution 5.22 Taking the derivative of both sides of

$$X(\Omega) = \frac{1}{\sqrt{2\pi}} \sum_{n=-\infty}^{\infty} x[n] e^{-j\Omega n}$$

with respect to Ω, we get

$$\frac{dX(\Omega)}{d\Omega} = \frac{1}{\sqrt{2\pi}} \sum_{n=-\infty}^{\infty} -jn x[n] e^{-j\Omega n}$$

where multiplying both sides by j, we obtain

$$j\frac{dX(\Omega)}{d\Omega} = \frac{1}{\sqrt{2\pi}} \sum_{n=-\infty}^{\infty} n x[n] e^{-j\Omega n}$$

from which it is seen that

$$j\frac{dX(\Omega)}{d\Omega}$$

is the Fourier transform of $nx[n]$, i.e., we get the transform pair

$$nx[n] \overset{\text{FT}}{\leftrightarrow} j\frac{dX(\Omega)}{d\Omega}$$

Example 5.23 Show that

$$x[n]y[n] \overset{\text{FT}}{\leftrightarrow} \frac{1}{\sqrt{2\pi}} X(\Omega) \otimes Y(\Omega) \tag{5.91}$$

Solution 5.23 The periodic convolution $X(\Omega) \otimes Y(\Omega)$ can be calculated as

$$X(\Omega) \otimes Y(\Omega) = \int_{2\pi} X(\lambda) Y(\Omega - \lambda) d\lambda$$

where substituting

$$X(\lambda) = \frac{1}{\sqrt{2\pi}} \sum_{n} x[n] e^{-j\lambda n}$$

and

$$Y(\Omega - \lambda) = \frac{1}{\sqrt{2\pi}} \sum_{m} y[m] e^{-j(\Omega-\lambda)m}$$

we obtain

Table 5.2 Properties of the Fourier transform

$x[n] \overset{FT}{\leftrightarrow} X(\Omega) \quad y[n] \overset{FT}{\leftrightarrow} Y(\Omega)$	$ax[n] + by[n] \overset{FT}{\leftrightarrow} aX(\Omega) + bY(\Omega)$
$x[n - n_0] \overset{FT}{\leftrightarrow} e^{-j\Omega n_0} X(\Omega)$	$e^{j\Omega_0 n} x[n] \overset{FT}{\leftrightarrow} X(\Omega - \Omega_0)$
$x^*[n] \overset{FT}{\leftrightarrow} X^*(-\Omega)$	$x[-n] \overset{FT}{\leftrightarrow} X(-\Omega)$
$x[n] * y[n] \overset{FT}{\leftrightarrow} \sqrt{2\pi} X(\Omega) Y(\Omega)$	$x[n]y[n] \overset{FT}{\leftrightarrow} \frac{1}{\sqrt{2\pi}} X(\Omega) \otimes Y(\Omega)$
	where $\otimes$ denotes periodic convolution
$\sum_n \lvert x[n] \rvert^2 = \int_{2\pi} \lvert X(\Omega) \rvert^2 d\Omega$	$x^*[-n] \overset{FT}{\leftrightarrow} X^*(\Omega)$
$nx[n] \overset{FT}{\leftrightarrow} j\frac{dX(\Omega)}{d\Omega}$	$x\left[\frac{n}{a}\right] \overset{FT}{\leftrightarrow} X(a\Omega)$

$$X(\Omega) \otimes Y(\Omega) = \frac{1}{2\pi} \int_{2\pi} \sum_n x[n] e^{-j\lambda n} \sum_m y[m] e^{-j(\Omega - \lambda)m} d\lambda$$

which can be rearranged as

$$X(\Omega) \otimes Y(\Omega) = \frac{1}{2\pi} \sum_n \sum_m x[n]y[m] e^{-j\Omega m} \int_{2\pi} e^{-j\lambda(n-m)} d\lambda \qquad (5.92)$$

where using

$$\int_{2\pi} e^{-j\lambda(n-m)} d\lambda = \begin{cases} 2\pi & n = m \\ 0 & n \neq m \end{cases} \qquad (5.93)$$

we get

$$X(\Omega) \otimes Y(\Omega) = \sum_n \sum_n x[n]y[n] e^{-j\Omega n}$$

Thus, we showed that

$$x[n]y[n] \overset{FT}{\leftrightarrow} \frac{1}{\sqrt{2\pi}} X(\Omega) \otimes Y(\Omega) \qquad (5.94)$$

Summary

Let $x[n]$ be a periodic signal with period N such that

$$x[n] = x[n + mN] \qquad (5.95)$$

One period of $x[n]$ denoted by $x_{op}[n]$ is defined as

$$x_{\text{op}}[n] = \begin{cases} x[n] & 0 \leq n \leq N-1 \\ 0 & \text{otherwise} \end{cases} \tag{5.96}$$

from which we can write that

$$x[n] = \sum_{m=-\infty}^{\infty} x_{\text{op}}[n - mN] \tag{5.97}$$

The Fourier series coefficients of $x[n]$ are calculated using

$$X[k] = \frac{1}{\sqrt{N}} \sum_{n,N} x[n] e^{-jk\Omega_0 n} \quad \Omega_0 = \frac{2\pi}{N} \tag{5.98}$$

which can be used to obtain the Fourier series representation of $x[n]$ as

$$x[n] = \frac{1}{\sqrt{N}} \sum_{k,N} X[k] e^{jk\Omega_0 n} \tag{5.99}$$

$X[k]$ is a complex periodic digital signal, and it can be considered a vector consisting of complex numbers. Each complex number in $X[k]$ has a magnitude and phase. $x_{\text{op}}[n]$ equals to one period of $x[n]$. $x_{\text{op}}[n]$ is a non-periodic signal

Fourier transform for digital signals is calculated using

$$X_{\text{op}}(\Omega) = \frac{1}{\sqrt{2\pi}} \sum_{n=-\infty}^{\infty} x_{\text{op}}[n] e^{-j\Omega n} \tag{5.100}$$

for which the inverse Fourier transform is calculated using

$$x_{\text{op}}[n] = \frac{1}{\sqrt{2\pi}} \int_{2\pi} X_{\text{op}}(\Omega) e^{j\Omega n} d\Omega \tag{5.101}$$

The Fourier transform function $X_{\text{op}}(\Omega)$ obtained for digital signals is a periodic function, and its period equals 2π. The digital signal $X[k]$ can be obtained by sampling $X_{\text{op}}(\Omega)$ in the frequency domain. The sampling frequency can be chosen of $\Omega_0 = \frac{2\pi}{N}$. Thus, we can write that

$$X[k] = \sqrt{\frac{2\pi}{N}} X_{\text{op}}(\Omega) \Big|_{\Omega = k\Omega_0} \quad \Omega_0 = \frac{2\pi}{N} \tag{5.102}$$

Problems

1. Are the following signals periodic? If they are periodic, find their periods:

   (a) $x_1[n] = \cos\left(\frac{3\pi}{17}n + \frac{\pi}{3}\right)$
   (b) $x_2[n] = 2\sin\left(\frac{14\pi}{19}n\right) + \cos(n) + 1$
   (c) $x_3[n] = \sum_{m=-\infty}^{\infty} (-1)^m \delta[n - 3m]$

2. $x[n] = 1 + e^{-j\frac{2\pi}{3}n} \rightarrow |X(\Omega)| = ?, \angle X(\Omega) = ?$

3. $x[n] = \delta[n+1] + 2\delta[n-1] \rightarrow X(\Omega) = ?, \quad |X(\Omega)| = ?, \quad \angle X(\Omega) = ?$

4. Using $x[n]$ in (2), we obtain the periodic signal $y[n] = \sum_{l=-\infty}^{\infty} x[n - 4l]$:

   (a) What is the period of $y[n]$?
   (b) Find the Fourier series coefficients of $y[n]$.

5. For $x[n] = -\delta[n+2] + \delta[n+1] + \delta[n-1] - \delta[n-2]$, calculate $X(\Omega)$ and roughly draw $|X(\Omega)|$. If $x[n]$ is accepted as a filter, determine whether $x[n]$ is a low-pass filter or high-pass filter.

6. Show that $x_1[n] * x_2[n] \overset{FT}{\leftrightarrow} \sqrt{2\pi} X_1(\Omega) X_2(\Omega)$.

7. Let $x_{op}[n]$ be a non-periodic signal and $x[n]$ be a periodic signal with period N. Prove the following.

   (a) $\sum_{n=-\infty}^{\infty} |x_{op}[n]|^2 = \int_{2\pi} |X_{op}(\Omega)|^2 d\Omega$

   (b) $\sum_{n=0}^{N-1} |x[n]|^2 = \sum_{k=0}^{N-1} |X[k]|^2$

8. If $X(\Omega) = \frac{-\frac{1}{3}e^{-j\Omega} + 2}{1 + \frac{1}{6}e^{-j\Omega} - \frac{1}{6}e^{-j2\Omega}}$, determine whether $x[n]$ is a periodic signal or not. Find $x[n]$.

9. The relationship between input and output of a linear time invariant system is given by

   $$y[n-2] - 5y[n-1] + 6y[n] = 4x[n-1] + 8x[n]$$

   where $x[n]$ is the system input and $y[n]$ is the system output:

   (a) Find the frequency response of the system.
   (b) Find the impulse response of the system.

10. If $x[n] \overset{FT}{\leftrightarrow} X(\Omega)$, then find the Fourier transforms of $nx[n]$ and $nx[n - n_0]$ in terms of $X(\Omega)$.

11. Find the Fourier transforms of the following signals:

(a) $x[n] = \left(\frac{1}{2}\right)^n u[n]$

(b) $x[n] = n\left(\frac{1}{2}\right)^n u[n-3]$

(c) $x[n] = n\left(\frac{1}{2}\right)^n e^{j\frac{\pi}{8}n} u[n-3]$

(d) $x[n] = \frac{\sin(\pi(n+1))}{\pi(n+1)}$

(e) $x[n] = \sin\left(\frac{\pi}{2}n\right)\left(\frac{1}{3}\right)^n u[n-3]$

(f) $x[n] = \left[\frac{\sin\left(\frac{\pi}{3}n\right)}{\pi n}\right]^3 * \frac{\sin\left(\frac{\pi}{3}n\right)}{\pi n}$

(g) $x[n] = \frac{\sin\left(\frac{\pi}{2}n\right)}{\pi n}$

(h) $x[n] = n\left(\frac{3}{5}\right)^{|n|}$

12. Find the digital periodic signals whose Fourier series coefficients are given as:

(a) $X[k] = \cos\left(\frac{\pi}{3}k\right)$

(b) $X[k] = j\delta[k-1] - j\delta[k+1]$ $\Omega_0 = 2\pi$

(c) $X[k] = \cos\left(\frac{4\pi}{21}k\right)$

13. Find the inverse Fourier transform of the following signals:

(a) $X(\Omega) = \cos(2\Omega) + j\sin(2\Omega)$

(b) $X(\Omega) = 2\cos(4\Omega)$

(c) $X(\Omega) = \sin(2\Omega) + \cos\left(\frac{\Omega}{2}\right)$

(d) $X(\Omega) = \frac{3}{-3 - 2^{-j\Omega} + e^{-j2\Omega}}$

14. The relationship between input and output of a linear time invariant system is given by

$$y[n-2] + 7y[n-1] + 6y[n] = 8x[n-1] + 10x[n-2]$$

Find the impulse response of this system.

15. $x[n] = (-1)^n$, $h[n] = \frac{\sin\left(\frac{\pi}{3}n\right)}{\pi n} \rightarrow y[n] = x[n] * h[n]$, $y[n] = ?$

16. $x[n] = (-1)^{n+1}$, $h[n] = \delta[n] - \frac{\sin\left(\frac{\pi}{3}n\right)}{\pi n} \rightarrow y[n] = x[n] * h[n]$, $y[n] = ?$

17. The relationships between the Fourier transforms of $x[n]$, i.e., $X(\Omega)$, and $y[n]$, i.e., $Y(\Omega)$ are given as:

(a) $Y(\Omega) = e^{-j2\Omega}X(\Omega)$

(b) $Y(\Omega) = Re\{X(\Omega)\}$

(c) $Y(\Omega) = \frac{d^2 X(\Omega)}{d\Omega^2}$

Express $y[n]$ in terms of $x[n]$ for each case.

18. Find the Fourier transform of $x[n] = \sin\left(\frac{\pi}{2}n\right) + \sin\left(\frac{\pi}{3}n\right)$.

19. For periodic signal $x[n]$, show that if

$$x[n] \xrightarrow{\text{DTFSC}} X[k]$$

then we have

$$X[n] \xrightarrow{\text{DTFSC}} x[-k]$$

Chapter 6
Laplace Transform

The Laplace transform was developed by the French mathematician and astronomer Pierre-Simon Laplace, who lived between 1749 and 1827. He has been called the Newton of France by some scientists.

The Laplace transform is a more general version of the continuous-time Fourier transform, and it can be applied to a broader class of signals. The Fourier transform is available for absolutely integrable signals. The absolute integral of an impulse response of an unstable system is not finite. It is not possible to examine the impulse response of such a system with the Fourier transform. However, the impulse response of the same system can be studied with the Laplace transform. Many properties of the Laplace transform are similar to the properties of Fourier transform.

There are two types of Laplace transform, one-sided and two-sided Laplace transforms. One-sided Laplace transform is used by real-time systems. The two-sided Laplace transform, on the other hand, does not make much sense for real-time systems, even if it is defined mathematically. Many systems can be easily analyzed, or many problems can be easily solved using the Laplace transform. Some of the areas where Laplace transform is applied are circuit analysis, solution of differential equations, root locus analysis in control engineering, etc. The output of a system can be calculated by multiplying the Laplace transform of the system's transfer function by the Laplace transform of its input. Using the Laplace transform of the impulse response of a linear and time-invariant system, we can determine many properties of the system.

In the following sections, we will first define the Laplace transform. Finding the convergence region of the Laplace transform is the second topic we will cover in this chapter. In sequel, properties of the Laplace transform will be covered. Laplace transforms of derivatives and integrals of functions and solutions of differential equations using Laplace transform are covered next. The Laplace transforms of some special functions are the last topic we will cover.

O. Gazi, *Principles of Signals and Systems*, https://doi.org/10.1007/978-3-031-17789-7_6

6.1   Laplace Transform

The Laplace transform is used for continuous-time signals, and it is a more general form of the Fourier transform. Laplace transform can be calculated for some signals whose Fourier transform is not available. For any continuous-time signal $f(t)$, two types of Laplace transform can be defined, one-sided and two-sided Laplace transforms. One-sided Laplace transform is defined as

$$F(s) = \int_{t=0}^{\infty} f(t)e^{-st}dt \qquad (6.1)$$

and two-sided Laplace transform is defined as

$$F(s) = \int_{t=-\infty}^{\infty} f(t)e^{-st}dt \qquad (6.2)$$

where s is a complex number defined as $s = \sigma + jw$. There is a close relationship between the Laplace transform and the Fourier transform. If the Fourier transform is defined as

$$\widehat{f}(w) = \frac{1}{\sqrt{2\pi}} \int_{t=-\infty}^{\infty} f(t)e^{-jwt}dt \qquad (6.3)$$

then the Laplace transform should be defined as

$$F(s) = \frac{1}{\sqrt{2\pi}} \int_{t=-\infty}^{\infty} f(t)e^{-st}dt \qquad (6.4)$$

Since many authors use the Laplace transform definition in their books as

$$F(s) = \int_{t=-\infty}^{\infty} f(t)e^{-st}dt \qquad (6.5)$$

we will continue this conventional definition throughout the chapter. In fact, the one-sided Laplace transform of $f(t)$ is nothing, but the two-sided Laplace transform of $f(t)u(t)$. We will use the two-sided Laplace transform when we refer to the Laplace transform unless otherwise stated in this book.

If the parameter s in the Laplace transform formula

$$F(s) = \int_{t=-\infty}^{\infty} f(t)u(t)e^{-st}dt \qquad (6.6)$$

is a purely imaginary complex number, i.e., if $s = jw$, then we get

$$F(s) = \int_{t=-\infty}^{\infty} f(t)u(t)e^{-jwt}dt \qquad (6.7)$$

which is Fourier transform formula. It is seen that the Laplace transform is a more generalized form of the Fourier transform.

Substituting $s = \sigma + jw$ in the Laplace transform formula

$$F(s) = \int_{t=-\infty}^{\infty} f(t)u(t)e^{-st}dt \qquad (6.8)$$

we get

$$F(\sigma + jw) = \int_{t=-\infty}^{\infty} f(t)u(t)e^{-(\sigma+jw)t}dt$$
$$= \int_{t=-\infty}^{\infty} f(t)u(t)e^{-\sigma t}e^{-jwt}dt \qquad (6.9)$$

from which it is seen that Laplace transform of $f(t)u(t)$ equals to the Fourier transform of $f(t)u(t)e^{-\sigma t}$. The magnitude of the product term $f(t)u(t)e^{-\sigma t}$ may increase of decrease depending on the value of σ.

The range of σ values, for which the integral

$$F(s) = \int_{t=-\infty}^{\infty} f(t)u(t)e^{-st}dt \qquad (6.10)$$

results in a finite value, is called region of convergence for the Laplace transform of $f(t)$. For some functions, the Laplace transform may not be found. For a function to have a Laplace transform in a certain region, it must satisfy some conditions. Let us now explain these conditions.

Theorem For $f(t)$ to have a one-sided Laplace transform, it should satisfy the following:

(a) $f(t)$ must be a piecewise continuous function on the interval $[0\cdots\infty)$.

(b) For $\text{Re}\{s\} > k$ on a region, the function must satisfy $|f(t)| \leq Ce^{kt}$.

Proof If the function $f(t)$ is piecewise continuous on the interval $(-\infty\cdots\infty)$, then the function $e^{-st}f(t)$ can be integrated on the interval $[0\cdots\infty)$. We have

$$|F(s)| = \left| \int_{t=0f(t)e^{-st}dt}^{\infty} \right| \leq \int_{t=0}^{\infty} |f(t)||e^{-st}|dt \leq \int_{t=0}^{\infty} Ce^{kt}e^{-\sigma t}dt = \frac{C}{k-\sigma}e^{-(\sigma-k)t}\big|_0^{\infty}$$

which for $\text{Re}\{s\} > k$ leads to

$$|F(s)| = \frac{C}{\sigma-k} < \infty$$

Example 6.1 Find the Laplace transform of $x(t) = e^{-at}u(t)$.

Solution 6.1 We can calculate the Laplace transform using

$$X(s) = \int_{t=-\infty}^{\infty} x(t)e^{-st}dt \tag{6.11}$$

where substituting $x(t) = e^{-at}u(t)$ we get

$$X(s) = \int_{t=-\infty}^{\infty} e^{-at}u(t)e^{-st}dt \tag{6.12}$$

leading to

$$X(s) = \int_{t=0}^{\infty} e^{-(s+a)t}dt \tag{6.13}$$

where substituting $s = \sigma + jw$, we obtain

$$X(s) = \int_{t=0}^{\infty} e^{-(\sigma+a+jw)t}dt$$

$$= \int_{t=0}^{\infty} e^{-(\sigma+a)t}e^{-jwt}dt \tag{6.14}$$

where e^{-jwt} denotes the unit circle and $|e^{-jwt}| = 1$. Then, the product term

$$\left| e^{-(\sigma+a)t} e^{-jwt} \right|$$

can be simplified as

$$\left| e^{-(\sigma+a)t} e^{-jwt} \right| = \left| e^{-(\sigma+a)t} \right| \left| e^{-jwt} \right| \rightarrow \left| e^{-(\sigma+a)t} \right|$$

For the integral in (6.14) to have a finite value,

$$e^{-(\sigma+a)t} \tag{6.15}$$

must be a decreasing function. For $e^{-(\sigma+a)t}$ to be a decreasing function, we must have

$$\sigma + a > 0 \tag{6.16}$$

which implies that $\mathrm{Re}\{s\} > -a$. That is, for the convergence of the Laplace transform, the condition $\mathrm{Re}\{s\} > -a$ must be satisfied.

Thus, (6.14) with the convergence condition (6.16) can be calculated as

$$
\begin{aligned}
X(s) &= \int_{t=0}^{\infty} e^{-(s+a)t} dt \\
&= \frac{1}{s+a} \quad \mathrm{Re}\{s\} > -a
\end{aligned}
\tag{6.17}
$$

The Fourier transform of the given signal can be calculated as

$$
\begin{aligned}
X(w) &= \int_{t=-\infty}^{\infty} x(t) e^{-jwt} dt \\
&= \int_{t=-\infty}^{\infty} e^{-at} u(t) e^{-jwt} dt \\
&= \int_{t=0}^{\infty} e^{-at} e^{-jwt} dt \\
&= \int_{t=0}^{\infty} e^{-(a+jw)t} dt \\
&= \frac{1}{a+jw} \quad a > 0
\end{aligned}
$$

Thus, we see that the criteria for the existence of Laplace transform is $\mathrm{Re}\{s\} > -a$ and the criteria for the existence of Fourier transform is $a > 0$.

If a is a negative number, the Fourier transform cannot be calculated, whereas the Laplace transform can be calculated since the Laplace transform condition can be satisfied for some σ values. That is, some signals may not have Fourier transforms, but the same signals may have Laplace transforms.

6.1.1   Convergence Region

In the previous section, we have seen that the condition for $x(t) = e^{-at}u(t)$ to have Laplace transform is that $\text{Re}\{s\} > -a$. Since s denotes a complex number, the inequality $\text{Re}\{s\} > -a$ indicates a region in the complex plane. This region is called the convergence region of the Laplace transform, and Laplace transform has finite value for the complex numbers falling into the convergence region.

For the regions outside the region of convergence, the Laplace integral diverges. The region described by the inequality $\text{Re}\{s\} > -a$ for a positive number a is depicted in Fig. 6.1.

Example 6.2   Calculate the Laplace transform of $x(t) = -e^{-at}u(-t)$.

Solution 6.2   Employing the Laplace transform formula

$$X(s) = \int_{t=-\infty}^{\infty} x(t)e^{-st}dt \qquad (6.18)$$

for the given signal, we get

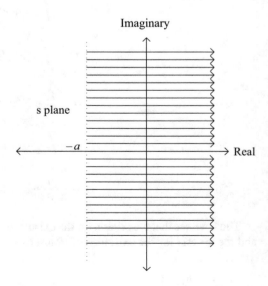

Fig. 6.1   The convergence region of the Laplace transform of $x(t) = e^{-at}u(t)$

$$X(s) = \int_{t=-\infty}^{\infty} -e^{-at}u(-t)e^{-st}dt \qquad (6.19)$$

leading to

$$X(s) = -\int_{t=-\infty}^{0} e^{-(s+a)t}dt \qquad (6.20)$$

where the integral gets finite values if $\text{Re}\{s+a\} < 0 \rightarrow \text{Re}\{s\} < -a$, and under this constraint, the integral can be evaluated as

$$\begin{aligned} X(s) &= -\int_{t=-\infty}^{0} e^{-(s+a)t}dt \\ &= \frac{1}{s+a} \end{aligned} \qquad (6.21)$$

We can express the obtained results in transform pairs as

$$\begin{aligned} e^{-at}u(t) &\overset{\mathcal{L}}{\leftrightarrow} \frac{1}{s+a} \qquad \text{Re}\{s\} > -a \\ -e^{-at}u(-t) &\overset{\mathcal{L}}{\leftrightarrow} \frac{1}{s+a} \qquad \text{Re}\{s\} < -a \end{aligned} \qquad (6.22)$$

Example 6.3 Find the Laplace transform of $f(t) = u(t)$.

Solution 6.3 Employing the Laplace transform formula, we get

$$\begin{aligned} F(s) &= \int_{t=-\infty}^{\infty} f(t)e^{-st}dt \\ &= \int_{t=-\infty}^{\infty} u(t)e^{-st}dt \\ &= \int_{t=0}^{\infty} e^{-st}dt \end{aligned}$$

where the integral gets a finite value if $\text{Re}\{s\} > 0$, and with this constraint the integral can be evaluated as

$$F(s) = \int_{t=0}^{\infty} e^{-st} dt$$
$$= \frac{1}{s}, \quad \text{Re}\{s\} > 0 \tag{6.23}$$

Example 6.4 Find the Laplace transform of $f(t) = u(t - a)$.

Solution 6.4 Employing the Laplace transform formula, we get

$$F(s) = \int_{t=-\infty}^{\infty} u(t-a) e^{-st} dt$$

$$= \int_{t=a}^{\infty} e^{-st} dt$$

$$= \left(-\frac{e^{-st}}{s} \Big|_{t=a}^{\infty} \right.$$

$$= \frac{e^{-as}}{s}, \quad \text{Re}\{s\} > 0$$

The obtained result can be expressed in a transform pair as

$$u(t - a) \overset{L}{\longleftrightarrow} \frac{e^{-as}}{s} \tag{6.24}$$

Note $L\{f(t - a)\} = e^{-as} F(s)$

Theorem Laplace transform is a linear operation. That is, we have

$$L\{af(t) + bg(t)\} = aF(s) + bG(s)$$
$$\mathcal{L}^{-1}\{aF(s) + bG(s)\} = af(t) + bg(t) \tag{6.25}$$

where $\mathcal{L}$ and $\mathcal{L}^{-1}$ indicate the Laplace and inverse Laplace transforms.

Note $f(t)$ If the convergence region for the Laplace transform of $f(t)$ is R_1 and the convergence for the Laplace transform of $g(t)$ is R_2, then the convergence region for the Laplace transform of $af(t) + bg(t)$ is $R_1 \cap R_2$.

Note Some authors accept the s variable in

$$F(s) = \int_{t=-\infty}^{\infty} f(t) e^{-st} dt \tag{6.26}$$

as a variable taking real values, and in this case, the convergence region of $x(t) = e^{-at}u(t)$ is indicated as $s > -a$ different than $Re\{s\} > -a$ used for complex s parameter.

Example 6.5 $f(t) = e^{-t}u(t) + e^{-3t}u(t)$, $F(s) = ?$

Solution 6.5 Previously, we obtained that

$$
\begin{aligned}
e^{-t}u(t) &\overset{\mathcal{L}}{\leftrightarrow} \frac{1}{s+1} \quad Re\{s\} > -1 \\
e^{-3t}u(t) &\overset{\mathcal{L}}{\leftrightarrow} \frac{1}{s+3} \quad Re\{s\} > -3
\end{aligned}
\tag{6.27}
$$

The convergence region of $F(s)$ is obtained as

$$
\{Re\{s\} > -1\} \cap \{Re\{s\} > -3\} = Re\{s\} > -1 \tag{6.28}
$$

Hence we can write that

$$
F(s) = \frac{1}{s+1} + \frac{1}{s+3} \quad Re\{s\} > -1 \tag{6.29}
$$

Example 6.6 $f(t) = \cosh(at)u(t)$, $a > 0$, $F(s) = ?$

Solution 6.6 Previously it is found that

$$
e^{-at}u(t) \overset{\mathcal{L}}{\leftrightarrow} \frac{1}{s+a}, \quad Re\{s\} > -a \tag{6.30}
$$

Hyperbolic cosine function can be written in terms of exponential signals as

$$
\begin{aligned}
f(t) &= \cosh(at)u(t) \\
&= \frac{e^{at} + e^{-at}}{2} u(t)
\end{aligned}
\tag{6.31}
$$

where taking the Laplace transform of both sides, we get

$$
\begin{aligned}
F(t) &= \frac{\mathcal{L}\{e^{at}u(t)\} + \mathcal{L}\{e^{-at}u(t)\}}{2} \\
&= \frac{\left(\frac{1}{s-a} + \frac{1}{s+a}\right)}{2}
\end{aligned}
$$

leading to

$$F(s) = \frac{s}{s^2 - a^2}, \quad \text{Re}\{s\} > a \tag{6.32}$$

The result can be expressed in a transform pair as

$$\cosh(at)u(t) \overset{\mathcal{L}}{\leftrightarrow} \frac{s}{s^2 - a^2} \quad \text{Re}\{s\} > -a \tag{6.33}$$

Example 6.7 $x(t) = e^{jat}u(t)$, $X(s) = ?$

Solution 6.7 The Laplace transform of $x(t) = e^{jat}u(t)$ can be calculated as

$$\begin{aligned} X(s) &= \mathcal{L}\{e^{jat}u(t)\} \\ &= \frac{1}{s - ja} \\ &= \frac{s + ja}{s^2 + a^2} \\ &= \frac{s}{s^2 + a^2} + j\frac{a}{s^2 + a^2} \quad \text{Re}\{s\} > 0 \end{aligned}$$

Example 6.8 Find the Laplace transforms of:

(a) $f(t) = \sin(at)$, $t > 0$ (b) $g(t) = \cos(at)$, $t > 0$

Solution 6.8 Taking the Laplace transform of both sides of

$$e^{jat} = \cos(at) + j\sin(at), t > 0$$

we get

$$\mathcal{L}\{e^{jat}\} = \mathcal{L}\{\cos(at)\} + j\mathcal{L}\{\sin(at)\}, \quad t > 0 \tag{6.34}$$

where using the result of the previous example

$$\mathcal{L}\{e^{jat}u(t)\} = \frac{s}{s^2 + a^2} + j\frac{a}{s^2 + a^2} \quad \text{Re}\{s\} > 0 \tag{6.35}$$

we get

$$\frac{s}{s^2 + a^2} + j\frac{a}{s^2 + a^2} = \mathcal{L}\{\cos(at)u(t)\} + j\mathcal{L}\{\sin(at)u(t)\} \tag{6.36}$$

from which we obtain

$$\cos{(at)}u(t) \overset{\mathcal{L}}{\leftrightarrow} \frac{s}{s^2+a^2}, \quad \text{Re}\{s\} > 0$$

$$\sin{(at)}u(t) \overset{\mathcal{L}}{\leftrightarrow} \frac{s}{s^2+a^2}, \quad \text{Re}\{s\} > 0$$

(6.37)

Theorem If the Laplace transform of $f(t)$ is $F(s)$, Re $\{s\} > k$, then the Laplace transform of $e^{at}f(t)$ is

$$F(s-a) \text{ with Re}\{s\} > \text{Re}\{a\} + k$$

That is, we have

$$f(t) \overset{\mathcal{L}}{\leftrightarrow} F(s)$$
$$\mathcal{L}\{e^{at}f(t)\} = F(s-a)$$
$$\mathcal{L}^{-1}\{F(s-a)\} = e^{at}f(t)$$

(6.38)

Proof Let $g(t) = e^{at}f(t)$; the Laplace transform of $g(t)$ can be calculated as

$$
\begin{aligned}
G(s) &= \int_{t=-\infty}^{\infty} g(t)e^{-st}dt \\
&= \int_{t=-\infty}^{\infty} e^{at}f(t)e^{-st}dt
\end{aligned}
$$

(6.39)

leading to

$$G(s) = \int_{t=-\infty}^{\infty} f(t)e^{-t(s-a)}dt$$

(6.40)

which is $F(s-a)$. The convergence region of (6.40) can be written as

$$\text{Re}\{s-a\} > k$$

(6.41)

from which we obtain Re$\{s\} > $ Re $\{a\} + k$.

Example 6.9 $\mathcal{L}\{e^{ct}\cos{(at)}u(t)\} = ?$

Solution 6.9 Using the shifting property of the Laplace transform for

$$\cos{(at)}u(t) \overset{\mathcal{L}}{\leftrightarrow} \frac{s}{s^2+a^2}, \text{Re}\{s\} > 0$$

we get

$$e^{ct} \cos{(at)}u(t) \overset{\mathcal{L}}{\leftrightarrow} \frac{s-c}{(s-c)^2+a^2}, \text{Re}\{s\}>c \qquad (6.42)$$

Example 6.10 $\mathcal{L}\{e^{ct} \sin{(at)}u(t)\} = ?$

Solution 6.10 Following a similar as in the previous example, we obtain

$$\mathcal{L}\{e^{ct} \sin{(at)}u(t)\} = \frac{c}{(s-c)^2+a^2}, \text{Re}\{s\}>c \qquad (6.43)$$

Note The Laplace transform of $\cos(at)u(t)$ equals to the one-sided Laplace transform of $\cos(at)$.

Example 6.11 Calculate the one-sided Laplace transform of $f(t) = e^{-at}$.

Solution 6.11 Previously, we have calculated the Laplace transform of $f(t) = e^{-at}u(t)$, and we obtained

$$e^{-at}u(t) \overset{\mathcal{L}}{\leftrightarrow} \frac{1}{s+a}, \quad \text{Re}\{s\} > -a \qquad (6.44)$$

This is the same question asked in a different way.

Example 6.12 If $f(t) = t^n u(t)$, then show that

$$F(s) = \frac{n!}{s^{n+1}} \qquad (6.45)$$

Solution 6.12 We can calculate the Laplace transform using

$$F(s) = \int_{t=0}^{\infty} t^n e^{-st} dt$$

where integral will be calculated for the given function using integration by parts rule

$$\int u\,dv = uv - \int v\,du$$

for which employing $u = t^n$ and $dv = e^{-st}dt$, we obtain

$$\mathcal{L}\{t^n\} = \int_0^\infty t^n e^{-st} dt$$

$$= \left(-\frac{1}{s} e^{-st} t^n \Big|_0^\infty - \int_{t=0}^\infty -\frac{n}{s} e^{-st} t^{n-1} dt \right)$$

$$= 0 + \frac{n}{s} \mathcal{L}\{t^{n-1}\}$$

$$\cdot$$
$$\cdot$$

$$= \frac{n!}{s^{n+1}}$$

Example 6.13 $f(t) = t^a u(t)$, $F(s) = ?$

Solution 6.13 Using Laplace transform formula, we get

$$\mathcal{L}\{t^a u(t)\} = \int_{t=0}^\infty t^a e^{-st} dt \tag{6.46}$$

where making use of the parameter change $x = st$, we obtain

$$\mathcal{L}\{t^a u(t)\} = \int_{t=0}^\infty \frac{1}{s} \left(\frac{x}{s}\right)^a e^{-x} dx \tag{6.47}$$

leading to

$$\mathcal{L}\{t^a u(t)\} = \frac{1}{s^{a+1}} \int_{x=0}^\infty x^a e^{-x} dx \tag{6.48}$$

which cannot be further simplified. The integral term in (6.48) can be expressed in terms of the gamma function

$$\Gamma(x) = \int_{y=0}^\infty e^{-y} y^{x-1} dy \tag{6.49}$$

as

$$\mathcal{L}\{t^a u(t)\} = \frac{1}{s^{a+1}}\Gamma(a+1) \tag{6.50}$$

Examples 6.14 Using the Laplace transform,

$$\mathcal{L}\{t^n u(t)\} = \frac{n!}{s^{n+1}} \tag{6.51}$$

and the property

$$\mathcal{L}\{f(t)e^{at}\} = F(s-a) \tag{6.52}$$

we can evaluate the following Laplace transforms:

(a) $\mathcal{L}\{tu(t)\} = \frac{1}{s^2}$ (b) $\mathcal{L}\{t^n e^{at} u(t)\} = \frac{n!}{(s-a)^{n+1}}$ (c) $\mathcal{L}\{te^{at} u(t)\} = \frac{1}{(s-a)^2}$

Properties of Laplace Transform
We have

$$f(t) \overset{\mathcal{L}}{\leftrightarrow} F(s), \ \text{ROC} = R \tag{6.53}$$

where ROC indicates the region of convergence. We have the properties:

Shifting in Time Domain

$$f(t - t_0) \overset{\mathcal{L}}{\leftrightarrow} e^{-st_0} F(s), \ \text{ROC} = R \tag{6.54}$$

where it is seen that shifting in time domain does not change the region of convergence.

Time Scaling

$$f(kt) \overset{\mathcal{L}}{\leftrightarrow} \frac{1}{|k|} F\left(\frac{s}{k}\right), \ \text{ROC} = \frac{R}{k} \tag{6.55}$$

where it is seen that scaling changes the region of convergence.

Convolution

$$f(t) * g(t) \overset{\mathcal{L}}{\leftrightarrow} F(s)G(s), \ \text{ROC} \supset (\text{ROC}_1 \cap \text{ROC}_2) \tag{6.56}$$

where ROC_1 and ROC_2 are the regions of convergences for $F(s)$ and $G(s)$.

Derivative in s-Plane

$$f(t) \overset{\mathcal{L}}{\leftrightarrow} -\frac{dF(s)}{ds}, \quad \text{ROC} = R \tag{6.57}$$

Derivative in the Time Domain

$$\frac{df(t)}{dt} \overset{\mathcal{L}}{\leftrightarrow} sF(s), \quad \text{ROC} \supset R \tag{6.58}$$

In these properties two-sided Laplace transform is considered; for one-sided Laplace transform, the properties show some differences. The properties for one-sided Laplace transform will be considered in the incoming sections.

Integration in the Time Domain

$$\int_{\tau=-\infty}^{t} f(\tau)d\tau \overset{\mathcal{L}}{\leftrightarrow} \frac{1}{s}F(s), \quad \text{ROC} \supset [R \cap \{\text{Re}\,\{s\} > 0\}] \tag{6.59}$$

Initial and Final Value Theorems
For causal signal $f(t)$, i.e., $f(t) = 0$, $t < 0$, we have the initial value theorem

$$f(0^+) = \lim_{s \to \infty} sF(s) \tag{6.60}$$

and final value theorem

$$\lim_{t \to \infty} f(t) = \lim_{s \to 0} sF(s) \tag{6.61}$$

Properties and Determination of Convergence Regions for Laplace Transforms
Before giving detailed information about the convergence region, let us explain the subject with an example.

Example 6.15 Calculate the Laplace transform of $f(t) = e^{-t}u(t) + e^{-4t}u(t)$, and determine the convergence region.

Solution 6.15 Using

$$
\begin{aligned}
e^{-t}u(t) &\overset{\mathcal{L}}{\leftrightarrow} \frac{1}{s+1}, \quad \text{Re}\,\{s\} > -1 \\
e^{-4t}u(t) &\overset{\mathcal{L}}{\leftrightarrow} \frac{1}{s+4}, \quad \text{Re}\,\{s\} > -4
\end{aligned} \tag{6.62}
$$

we can calculate the Laplace transform of the given signal as

$$F(s) = \frac{1}{s+1} + \frac{1}{s+4}$$
$$= \frac{2s+5}{s^2+5s+4}, \quad \text{Re}\{s\} > -1$$

It is seen from the solution of this example that the Laplace transform is a mathematical term in the form $\frac{N(s)}{D(s)}$ and the boundaries of the convergence region are related to the poles of the $\frac{N(s)}{D(s)}$ rational expression.

The graph containing the locations of the zeros and poles of $\frac{N(s)}{D(s)}$ is called pole-zero graph. Let us give some properties of the convergence region and prove some of these properties later on.

Right-Sided Function
If $f(t) = 0$, $t < t_0$ and $f(t) \neq 0$, $t > t_0$, then $f(t)$ is called a right-sided function.

Left-Sided Function
If $f(t) = 0$, $t > t_0$ and $f(t) \neq 0$, $t < t_0$, then $f(t)$ is called a left-sided function.

Two-Sided Function
If $f(t) = 0$ for $t < t_0$ or $t > t_1$ and $f(t) \neq 0$ for $t_0 < t < t_1$, then $f(t)$ is called a two-sided function.

Property 1 The convergence region contains stripes parallel to the vertical axis.

Property 2 For rational Laplace transform functions $\frac{N(s)}{D(s)}$, the convergence region does not contain any poles of the transfer fuction.

Property 3 If $f(t)$ is a right-sided function, the region of convergence of the Laplace transform is denoted by $\text{Re}\{s\} > \sigma_0$, and for rational Laplace transform functions, the region of convergence is to the right side of the rightmost pole in the s plane.

Property 4 If $f(t)$ is a left-sided function, the region of convergence of the Laplace transform is denoted by $\text{Re}\{s\} < \sigma_0$, and for rational Laplace transform functions, the region of convergence is to the left side of the leftmost pole in the S plane.

Property 5 If $f(t)$ is a two-sided function, the region of convergence of the Laplace transform is denoted by $\sigma_0 < \text{Re}\{s\} < \sigma_1$, and for rational Laplace transform functions, the region of convergence is the zone between the two innermost poles in the s-plane.

Property 6 If the convergence region contains the imaginary axis, in addition to the Laplace transform, the Fourier transform can also be calculated. If the convergence region does not contain the imaginary axis, there is no Fourier transform, although there is a Laplace transform.

Property 7 For Laplace transform polynomials, the plane of convergence is the entire s-plane if there is at least one number in s-plane for which the Laplace transform converges.

Example 6.16 Find the Laplace transform of $f(t) = e^{-\alpha|t|}$, and determine the region of convergence.

Solution 6.16 Using the definition of absolute value, the function

$$f(t) = e^{-\alpha|t|} \tag{6.63}$$

can be written as

$$f(t) = \begin{cases} e^{-at} & t > 0 \\ e^{at} & t < 0 \end{cases} \tag{6.64}$$

which can be written in terms of unit step functions as

$$f(t) = e^{-at}u(t) + e^{at}u(-t) \tag{6.65}$$

Using the transform pairs

$$\begin{aligned} e^{-at}u(t) &\overset{\mathcal{L}}{\leftrightarrow} \frac{1}{s+a} \quad \text{Re}\{s\} > -a \\ e^{at}u(-t) &\overset{\mathcal{L}}{\leftrightarrow} \frac{-1}{s-a} \quad \text{Re}\{s\} < a \end{aligned} \tag{6.66}$$

the Laplace transform of (6.65) can be calculated as

$$\begin{aligned} F(s) &= \frac{1}{s+a} - \frac{1}{s-a} \\ &= \frac{2a}{s^2 - a^2}, \quad -a < \text{Re}\{s\} < a \end{aligned} \tag{6.67}$$

for which the convergence region is shown in Fig. 6.2.

Example 6.17 Determine the possible convergence regions for the Laplace transform

$$F(s) = \frac{1}{(s+1)(s-3)} \tag{6.68}$$

Solution 6.17 The poles of $F(s)$ are -1 and 3. The convergence regions can be selected as:

(a) $\text{Re}\{s\} < -1$ (b) $\text{Re}\{s\} > 3$ (c) $-1 < \text{Re}\{s\} < 3$ (d) Entire s-plane

Fig. 6.2 Laplace transform
convergence region of
$f(t) = e^{-a|t|}$

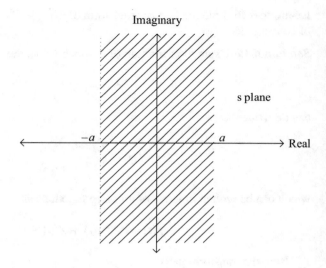

6.2   Inverse Laplace Transform

The inverse Laplace transform is defined as

$$f(t) = \frac{1}{2\pi j} \lim_{w \to \infty} \int_{\sigma - jw}^{\sigma + jw} F(s)e^{st}\,ds. \tag{6.69}$$

where $s = \sigma + jw$. The integration in (6.69) is a complex integration, and it is not
used in practice; instead, known Laplace transform pairs are used for the calculation
of inverse Laplace transforms.

Example 6.18   The Laplace transform of $f(t)$ is given as

$$F(s) = \frac{1}{(s + 1)(s + 3)}, \quad \text{Re}\{s\} > -1 \tag{6.70}$$

Find $f(t)$.

Solution 6.18   The Laplace transform expression can be written as the sum of two
rational polynomials as

$$\frac{1}{(s + 1)(s + 3)} = \frac{A}{s + 1} + \frac{B}{s + 3} \tag{6.71}$$

where $A = \frac{1}{2}$ and $B = -\frac{1}{2}$. Considering the convergence region indicated by
$\text{Re}\{s\} > -1$, and the right-hand side of (6.71), $f(t)$ can be written as

$$f(t) = \frac{1}{2}e^{-t}u(t) - \frac{1}{2}e^{-3t}u(t) \tag{6.72}$$

Example 6.19 Find the Laplace transform of $f(t) = e^{-at}(u(t) - u(t - T))$, and determine the convergence region.

Solution 6.19 Laplace transform can be calculated as

$$F(s) = \int_{t=-\infty}^{\infty} f(t)e^{-st}dt$$

$$= \int_{t=0}^{T} e^{-at}e^{-st}dt$$

leading to

$$F(s) = \frac{1}{s+a}\left(1 - e^{(s+a)T}\right)$$

where it seems at the first glance that $F(s)$ has a pole at $s = -a$; however, if we put $s = -a$ in $F(s)$, we get $\frac{0}{0}$, and the value of $F(s)$ at $s = -a$ can be calculated using L'Hospital's rule as

$$\lim_{s \to -a} F(s) = \lim_{s \to -a} \left[\frac{\dfrac{d\left(1 - e^{-(s+a)T}\right)}{ds}}{\dfrac{d(s+a)}{ds}}\right]$$

$$= \lim_{s \to -a} Te^{-sT}e^{-aT}$$

$$= T$$

Thus, we can conclude that the convergence region is the entire s-plane.

6.3 Laplace Transforms of Derivatives and Integrals of Functions

For the one-sided Laplace transforms of derivative and integral functions, the following properties are available. ROC means region of convergence

$$f(t) \overset{\mathcal{L}}{\leftrightarrow} F(s), \quad \text{ROC} = R \tag{6.73}$$

1.

$$L\{f'(t)\} = sF(s) - f(0), \text{ROC} = R \qquad (6.74)$$

2.

$$L\{f''(t)\} = s^2F(s) - sf(0) - f'(0), \ \text{ROC} = R \qquad (6.75)$$

3.

$$L\left\{\frac{df^3(t)}{dt^3}\right\} = s^3F(s) - s^2f(0) - sf'(0) - f''(0), \text{ROC} = R \qquad (6.76)$$

A more general expression from which the first three properties can be obtained is given in

4.

$$L\{f^n(t)\} = s^nF(s) - s^{n-1}f(0) - s^{n-2}f'(0) - \ldots - f^{n-1}(0), \text{ROC} = R \qquad (6.77)$$

Let us prove only the first property, the other properties can be proved in a similar manner.

Proof 1 The one-sided Laplace transform of the derivative function can be calculated using the property

$$\int u\,dv = uv - \int v\,du$$

as

$$L\{f'(t)\} = \int_{t=0}^{\infty} f'(t)e^{-st}dt, \ \ u = e^{-st}, dv = f'(t)dt$$

$$= (e^{-st}f(t)\Big|_0^{\infty} + s\int_0^{\infty} f(t)e^{-st}dt$$

$$= 0 - f(0) + sF(s)$$

resulting in

$$L\{f'(t)\} = sF(s) - f(0) \tag{6.78}$$

Example 6.20 Show that one-sided Laplace transform of $f(t) = \sin(t)$ is

$$F(s) = \frac{1}{s^2 + 1} \tag{6.79}$$

Solution 6.20 The first and second derivatives of $f(t) = \sin(t)$ are

$$f'(t) = \cos(t) \text{ ve } f''(t) = -\sin(t) \tag{6.80}$$

from which we can write that

$$f(t) = -f''(t) \tag{6.81}$$

Taking the one-sided Laplace transform of both sides of (6.81), we get

$$L\{f(t)\} = -L\{f''(t)\} \tag{6.82}$$

where using the property

$$L\{f''(t)\} = s^2 F(s) - sf(0) - f'(0)$$

we obtain

$$F(s) = -\left(s^2 F(s) - sf(0) - f'(0)\right) \tag{6.83}$$

resulting in

$$F(s) = -s^2 F(s) + s0 + 1 \tag{6.84}$$

from which we obtain

$$F(s) = \frac{1}{s^2 + 1} \tag{6.85}$$

Exercise Find the one-sided Laplace transform of $f(t) = t\sin(wt)$ using the property

$$L\{f''(t)\} = s^2 F(s) - sf(0) - f'(0) \tag{6.86}$$

Example 6.21 Find the solutions of the differential equation

$$y''(t) - y(t) = t, \ y(0) = 1, y'(0) = 1 \tag{6.87}$$

Solution 6.21 If we take the one-sided Laplace transform of both sides of

$$y''(t) - y(t) = t \tag{6.88}$$

we get

$$s^2 Y(s) - sy(0) - y'(0) - Y(s) = \frac{1}{s^2} \tag{6.89}$$

where employing the given initial values we obtain

$$Y(s) = \frac{s+1}{s^2-1} + \frac{1}{s^2(s^2-1)}$$

where using

$$\frac{s+1}{s^2-1} = \frac{1}{s-1}$$

and

$$\frac{1}{s^2(s^2-1)} = \frac{1}{s^2} + \frac{1}{s^2-1}$$

we get

$$Y(s) = \frac{1}{s-1} + \frac{1}{s^2-1} - \frac{1}{s^2} \tag{6.90}$$

Evaluating the inverse Laplace transform of (6.90), we obtain the time domain expression of the function as in

$$y(t) = \mathcal{L}^{-1}\left\{\frac{1}{s-1}\right\} + \mathcal{L}^{-1}\left\{\frac{1}{s^2-1}\right\} - \mathcal{L}^{-1}\left\{\frac{1}{s^2}\right\}$$
$$= e^t u(t) + \sinh(t)u(t) - tu(t)$$
$$= (e^t + \sinh(t) - t)u(t)$$

Note

$$\sinh(t) = \frac{e^t - e^{-t}}{2}, \cos(h) = \frac{e^t + e^{-t}}{2} \tag{6.91}$$

Note Since we study functions with known Laplace transforms in our examples, we will not specify convergence regions unless necessary.

Laplace Transforms of Integral Functions

Let us denote the one-sided Laplace transform of $f(t)$ by $F(s)$. The Laplace transform of integral of $f(t)$ can be calculated as

$$\mathcal{L}\left\{ \int_{\tau=0}^{t} f(\tau)d\tau \right\} = \frac{F(s)}{s} \tag{6.92}$$

Proof Let

$$g(t) = \int_{\tau=0}^{t} f(\tau)d\tau$$

which satisfy $g(0) = 0$, since the upper frontier of the integral is also zero for $t = 0$, i.e., we have

$$\int_{\tau=0}^{0} (\ldots)d\tau = 0$$

From

$$g(t) = \int_{\tau=0}^{t} f(\tau)d\tau$$

we get

$$f(t) = g'(t)$$

from which we obtain

$$\mathcal{L}\{g'(t)\} = sG(s) - g(0)$$

leading to

$$F(s) = sG(s)$$

which is rearranged as

$$G(s) = \frac{F(s)}{s}$$

Thus, we have

$$\mathcal{L}\left\{ \int_{\tau=0}^{t} f(\tau)d\tau \right\} = \frac{F(s)}{s} \tag{6.93}$$

Note

$$\mathcal{L}\left\{ \int_{\tau=0}^{t} f(\tau)d\tau \right\} = \frac{F(s)}{s} \Rightarrow \mathcal{L}^{-1}\left\{ \frac{F(s)}{s} \right\} = \int_{\tau=0}^{t} f(\tau)d\tau \tag{6.94}$$

Exercise Let $|f(t)| \le Me^{kt}$, define

$$g(t) = \int_{\tau=0}^{t} f(\tau)d\tau \tag{6.95}$$

whose absolute value satisfies

$$|g(t)| \le Ne^{kt}$$

Find the relationship between M and N.

Example 6.22 Laplace transform of $g(t)$ is given as

$$G(s) = \frac{1}{s(s^2 + w^2)} \tag{6.96}$$

Find $g(t)$.

Solution 6.22 It is known that

$$\mathcal{L}\{ \sin(wt)u(t) \} = \frac{w}{s^2 + w^2} \tag{6.97}$$

from which we obtain

$$\mathcal{L}\left\{ \frac{\sin(wt)u(t)}{w} \right\} = \frac{1}{s^2 + w^2} \tag{6.98}$$

which leads to

$$\mathcal{L}^{-1}\left\{\frac{1}{s^2 + w^2}\right\} = \frac{\sin{(wt)}u(t)}{w} \tag{6.99}$$

for which using the property

$$\mathcal{L}^{-1}\left\{\frac{F(s)}{s}\right\} = \int_{\tau=0}^{t} f(\tau)d\tau \tag{6.100}$$

where

$$F(s) = \frac{1}{s^2 + w^2} \tag{6.101}$$

and

$$f(\tau) = \frac{\sin{(w\tau)}}{w}u(\tau) \tag{6.102}$$

we obtain

$$\mathcal{L}^{-1}\left\{\frac{1}{s(s^2 + w^2)}\right\} = \int_{\tau=0}^{t} \frac{\sin{(w\tau)}}{w}d\tau = \frac{1}{w^2}(1 - \cos{(wt)})u(t) \tag{6.103}$$

Note The expression

$$\mathcal{L}^{-1}\left\{\frac{F(s)}{s}\right\} = \int_{\tau=0}^{t} f(\tau)d\tau \tag{6.104}$$

can be used in a recursive manner, i.e., we can write

$$\mathcal{L}^{-1}\left\{\frac{F(s)}{s^2}\right\} = \int_{0}^{t}\int_{0}^{\tau_2} f(\tau_1)d\tau_1 d\tau_2 \tag{6.105}$$

and

$$\mathcal{L}^{-1}\left\{\frac{F(s)}{s^3}\right\} = \int_{0}^{t}\int_{0}^{\tau_3}\int_{0}^{\tau_2} f(\tau_1)d\tau_1 d\tau_2 d\tau_3 \tag{6.106}$$

Example 6.23 $H(s)$ is given as

$$H(s) = \frac{1}{s^2(s^2 + w^2)} \tag{6.107}$$

Find $h(t)$.

Solution 6.23 In the previous example, for

$$G(s) = \frac{1}{s(s^2 + w^2)} \tag{6.108}$$

we found that

$$g(t) = \frac{1}{w^2}(1 - \cos(wt))u(t) \tag{6.109}$$

In this example, $H(s) = \frac{G(s)}{s}$ is given, and $h(t)$ is asked. Using the property

$$\mathcal{L}^{-1}\left\{\frac{G(s)}{s}\right\} = \int_{\tau=0}^{t} g(\tau)d\tau \tag{6.110}$$

where

$$h(t) = \mathcal{L}^{-1}\left\{\frac{G(s)}{s}\right\} \tag{6.111}$$

we obtain

$$\begin{aligned} h(t) &= \int_{\tau=0}^{t} \frac{1}{w^2}(1 - \cos(w\tau))d\tau \\ &= \frac{t}{w^2} - \frac{\sin(wt)}{w^3}, \quad t \geq 0 \end{aligned} \tag{6.112}$$

Example 6.24 Find the solution of the differential equation

$$y'' + y = 2t, \quad y\left(\frac{\pi}{4}\right) = \frac{\pi}{2}, \quad y'\left(\frac{\pi}{4}\right) = 2 - \sqrt{2} \tag{6.113}$$

Solution 6.24 Taking the one-sided Laplace transform of both sides of

$$y'' + y = 2t \tag{6.114}$$

we get

$$L\{y'' + y\} = L\{2t\} \rightarrow L\{y''\} + L\{y\} = 2L\{t\} \tag{6.115}$$

leading to

$$s^2 Y(s) - sy(0) - y'(0) + Y(s) = \frac{2}{s^2} \tag{6.116}$$

where for the terms $y(0)$ and $y'^{(0)}$ we can use constant numbers, i.e., $y(0) = c_1$, $y'(0) = c_2$. Then, we get

$$s^2 Y(s) - sc_1 - c_2 + Y(s) = \frac{2}{s^2} \tag{6.117}$$

which can be rearranged as

$$Y(s) = \frac{2}{s^2} - \frac{2}{s^2 + 1} + c_1 \frac{s}{s^2 + 1} + c_2 \frac{1}{s^2 + 1} \tag{6.118}$$

whose inverse Laplace transform can be calculated as

$$y(t) = 2t + c_1 \cos(t) + (c_2 - 2)\sin(t), \quad t \ge 0 \tag{6.119}$$

The constant terms c_1 and c_2 can be calculated using the equations

$$y\left(\frac{\pi}{4}\right) = \frac{\pi}{2}, \quad y'\left(\frac{\pi}{4}\right) = 2 - \sqrt{2} \tag{6.120}$$

as

$$y\left(\frac{\pi}{4}\right) = \frac{\pi}{2} \rightarrow \frac{\pi}{2} + \frac{c_1}{\sqrt{2}} + \frac{c_2 - 2}{\sqrt{2}} = \frac{\pi}{2}$$

$$y'\left(\frac{\pi}{4}\right) = 2 - \sqrt{2} \rightarrow 2 - \frac{c_1}{\sqrt{2}} + \frac{c_2 - 2}{\sqrt{2}} = 2 - \sqrt{2}$$

which can be simplified as

$$\frac{c_1}{\sqrt{2}} + \frac{c_2 - 2}{\sqrt{2}} = 0$$

$$\frac{c_1}{\sqrt{2}} - \frac{c_2 - 2}{\sqrt{2}} = \sqrt{2}$$

from which we find $c_1 = 1$, $c_2 = 1$. Hence, the solution can be written as

$$y(t) = 2t + \cos(t) - \sin(t), \quad t \geq 0 \tag{6.121}$$

Example 6.25 $f(t)$ is given as

$$f(t) = \frac{1}{2}t^2 u(t-1) \tag{6.122}$$

Find $F(s)$.

Solution 6.25 To find the Laplace transform of $f(t) = \frac{1}{2}t^2 u(t-1)$, we can use the properties

$$u(t-a) \overset{\mathcal{L}}{\leftrightarrow} \frac{e^{-as}}{s} \tag{6.123}$$

and

$$\mathcal{L}\{f(t-a)\} = e^{-as} F(s) \tag{6.124}$$

Substituting $t - 1 + 1$ for t in

$$\frac{1}{2}t^2 u(t-1)$$

we obtain

$$\begin{aligned} \frac{1}{2}t^2 u(t-1) &= \frac{1}{2}(t-1+1)^2 u(t-1) \\ &= \left(\frac{1}{2}(t-1)^2 + (t-1) + \frac{1}{2}\right)u(t-1) \\ &= \frac{1}{2}(t-1)^2 u(t-1) + (t-1)u(t-1) + \frac{1}{2}u(t-1) \end{aligned}$$

where taking the Laplace transform of both sides, we get

$$\mathcal{L}\left\{\frac{1}{2}t^2 u(t-1)\right\} = \mathcal{L}\left\{\frac{1}{2}(t-1)^2 u(t-1) + (t-1)u(t-1) + \frac{1}{2}u(t-1)\right\}$$

where using the properties

$$L\{t^2 u(t)\} = \frac{1}{s^3}, \quad L\{tu(t)\} = \frac{1}{s^2}, \quad L\{u(t)\} = \frac{1}{s}$$

we get

$$L\left\{\frac{1}{2}t^2 u(t-1)\right\} = \frac{e^{-s}}{s^3} + \frac{e^{-s}}{s^2} + \frac{e^{-s}}{2s}$$
$$= \left(\frac{1}{s^3} + \frac{1}{s^2} + \frac{1}{2s}\right)e^{-s}$$

Exercise $f(t)$ is given as

$$f(t) = \sin(t)u\left(t - \frac{\pi}{2}\right) \tag{6.125}$$

Find $F(s)$.

Example 6.26 $f(t)$ is given as

$$f(t) = \delta(t - a) \tag{6.126}$$

Find $F(s)$.

Solution 6.26 Using the property

$$\int \delta(t - a)f(t)dt = f(a) \tag{6.127}$$

we can calculate the Laplace transform of $f(t) = \delta(t - a)$ as

$$F(s) = \int_{t=0}^{\infty} \delta(t - a)e^{-st}dt \tag{6.128}$$
$$= e^{-as}$$

Exercise Find the solution of

$$y''(t) + 3y'(t) + 2y(t) = \delta(t - 2), \quad y(0) = 0, \quad y'(0) = 0 \tag{6.129}$$

using Laplace transform.

Property Let the convolution of the two functions $f(t)$ and $g(t)$ be $h(t) = f(t) * g(t)$. The Laplace transform of $h(t)$ can be written as

$$H(s) = F(s)G(s) \tag{6.130}$$

That is, we have the pair

$$f(t) * g(t) \overset{L}{\leftrightarrow} F(s)G(s) \qquad (6.131)$$

Proof The one-sided Laplace transform of $h(t)$ can be calculated using

$$H(s) = \int_{t=0}^{\infty} h(t)e^{-st}dt \qquad (6.132)$$

where substituting

$$h(t) = f(t) * g(t) = \int_{\tau=0}^{\infty} f(\tau)g(t-\tau)d\tau \qquad (6.133)$$

we obtain

$$H(s) = \int_{t=0}^{\infty} \left[\int_{\tau=0}^{t} f(\tau)g(t-\tau)d\tau \right] e^{-st}dt \qquad (6.134)$$

which is a two-dimensional integration, and it is evaluated for the region indicated by $0 \le t < \infty$ and $0 \le \tau \le t$ which is displayed in Fig. 6.3.

The region indicated by $0 \le t < \infty$ and $0 \le \tau \le t$ can also be referred to by the constraint $0 \le \tau < \infty$ and $\tau \le t \le \infty$, and then the integral in (6.134) can be written as

Fig. 6.3 The region indicated by $0 \le t < \infty$ and $0 \le \tau \le t$

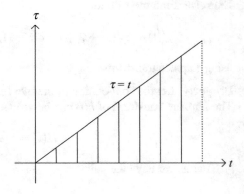

$$H(s) = \int\limits_{\tau=0}^{\infty} \int\limits_{t=\tau}^{\infty} f(\tau)g(t-\tau)e^{-st}dt d\tau \qquad (6.135)$$

where changing the parameters as

$$t' = t - \tau \rightarrow dt' = dt, \ \tau \geq 0 \ \ t' \geq 0 \qquad (6.136)$$

we obtain

$$H(s) = \int\limits_{\tau=0}^{\infty} \int\limits_{t'=0}^{\infty} f(\tau)g(t')e^{-s(t'+\tau)}dt' d\tau \qquad (6.137)$$

which can be rearranged as

$$H(s) = \underbrace{\int\limits_{\tau=0}^{\infty} f(\tau)e^{-s\tau}d\tau}_{F(s)} \underbrace{\int\limits_{t'=0}^{\infty} g(t')e^{-st'}dt'}_{G(s)} \qquad (6.138)$$

Hence, we showed that

$$f(t) * g(t) \overset{L}{\leftrightarrow} F(s)G(s) \qquad (6.139)$$

which is also valid for two-sided Laplace transform.

Property If $f(t) \overset{L}{\leftrightarrow} F(s)$, then we have $tf(t) \overset{L}{\leftrightarrow} -F'(s)$.

Proof Taking the derivative of

$$F(s) = \int\limits_{t=-\infty}^{\infty} f(t)e^{-st}dt \qquad (6.140)$$

with respect to s, we get

$$\frac{dF(s)}{ds} = -\int\limits_{t=-\infty}^{\infty} tf(t)e^{-st}dt \qquad (6.141)$$

from which we obtain

$$tf(t) \overset{\mathcal{L}}{\leftrightarrow} - F'(s) \tag{6.142}$$

Property If $f(t) \overset{\mathcal{L}}{\leftrightarrow} F(s)$, then we have

$$\frac{f(t)}{t} \overset{\mathcal{L}}{\leftrightarrow} \int\limits_{s'=s}^{\infty} F(s')ds' \tag{6.143}$$

Exercise $f(t) = \frac{1}{\sqrt{t}} \rightarrow F(s) = ?$

Solution of Bessel Equation Using Laplace Transform

The Bessel function is obtained by solving the Bessel equation. Bessel equation is given as

$$t^2 \frac{df^2(t)}{dt^2} + t\frac{df(t)}{dt} + (t^2 - c^2)f(t) = 0 \tag{6.144}$$

where c is a constant number. The Eq. (6.144) is a second-order equation and, there are two solutions to this equation. The two solutions of this equation are denoted by $J_\nu(t)$ and $Y_\nu(t)$. The first solution, $J_\nu(t)$, is called the first kind of Bessel function, and the second solution, $Y_\nu(t)$, is called the second kind of Bessel function. For $c = 0$, (6.144) takes the form

$$t\frac{df^2(t)}{dt^2} + \frac{df(t)}{dt} + tf(t) = 0$$

whose first solution is indicated by $J_0(t)$. Let us determine $J_0(t)$ for the initial conditions $J_0(0) = 1$ and $J'_0(0) = 1$.

The Laplace transform of the equation

$$t\frac{df^2(t)}{dt^2} + \frac{df(t)}{dt} + tf(t) = 0$$

can be calculated using the properties

$$\mathcal{L}\{f'(t)\} = sF(s) - f(0), \quad \mathcal{L}\{f''(t)\} = s^2 F(s) - sf(0) - f'(0)$$
$$tf(t) \overset{\mathcal{L}}{\leftrightarrow} - F'(s)$$

as

$$-\left(s^2 F(s)\right)' + sF(s) - F'(s) = 0$$

leading to

$$- 2sF(s) - s^2 F'(s) + sF(s) - F'(s) = 0$$

from which we get

$$F'(s) = - \frac{s}{s^2 + 1} F(s) \tag{6.145}$$

The solution of (6.145) can be evaluated as

$$\frac{dF(s)}{ds} = - \frac{s}{s^2 + 1} F(s) \rightarrow \frac{dF(s)}{F(s)} = - \frac{s}{s^2 + 1} ds$$

where taking the integral of both sides, we obtain

$$\int \frac{dF(s)}{F(s)} = - \int \frac{s}{s^2 + 1} ds \rightarrow \ln(F(s)) + k_1 = - \frac{1}{2} \ln(s^2 + 1) + k_2$$

resulting in

$$F(s) = \frac{C}{(s^2 + 1)^{\frac{1}{2}}}$$

where the constant C can be calculated using the initial value theorem

$$J_0(0) = \lim_{s \to \infty} sF(s)$$

where using $J_0(0) = 1$ we get

$$1 = \lim_{s \to \infty} \left(s \frac{C}{(s^2 + 1)^{\frac{1}{2}}} \right) = C$$

Thus, we found that

$$\mathcal{L}\{J_0(t)\} = F(s) = \frac{1}{(s^2 + 1)^{\frac{1}{2}}}, \quad s > 0 \tag{6.146}$$

To find $J_0(t)$, let us first write the Binomial expansion of

$$\mathcal{L}\{J_0(t)\} = \frac{1}{2} \left(1 + \frac{1}{s^2} \right)^{-\frac{1}{2}}$$

as in

$$\mathcal{L}\{J_0(t)\} = \frac{1}{s} - \frac{1}{2}\frac{1}{s^3} + \frac{3}{8}\frac{1}{s^5} - \frac{5}{16}\frac{1}{s^7} + \cdots$$

whose inverse Laplace transform can be written as

$$J_0(t) = 1 - \frac{t^2}{4} + \frac{t^4}{64} - \frac{t^6}{2304} + \cdots$$

which can be written using the summation symbol as

$$J_0(t) = \sum_{n=0}^{\infty} \frac{(-1)^n t^{2n}}{2^{2n}(n!)^2} \tag{6.147}$$

which is the solution of the Bessel equation.

Example 6.27 Find the Laplace transform of

$$\frac{\sin(t)}{t} u(t) \tag{6.148}$$

Solution 6.27 Using the transform pairs

$$\sin(t)u(t) \overset{\mathcal{L}}{\leftrightarrow} \frac{1}{s^2 + 1} \quad \text{Re}\{s\} > 0, \quad \frac{f(t)}{t} \overset{\mathcal{L}}{\leftrightarrow} \int_{s}^{\infty} F(s)\,ds$$

we can calculate the Laplace transform of

$$\frac{\sin(t)}{t} u(t)$$

as in

$$\frac{\overbrace{\mathrm{Sin}(t)}^{f(t)}}{t} u(t) \overset{\mathcal{L}}{\leftrightarrow} \int_{s}^{\infty} F(s)\,ds$$

where using

$$F(s) = \frac{1}{s^2 + 1}$$

we get

$$\int\limits_{s}^{\infty} F(s)ds = \int\limits_{s}^{\infty} \frac{ds}{s^2 + 1} \rightarrow \frac{1}{2}\left(\pi - 2\tan^{-1}(s)\right)$$

Thus, we got

$$\frac{\sin(t)}{t}u(t) \overset{\mathcal{L}}{\leftrightarrow} \frac{1}{2}\left(\pi - 2\tan^{-1}(s)\right) \tag{6.149}$$

Example 6.28 Find the Laplace transform of the signal depicted in Fig. 6.4.

Solution 6.28 The signal in Fig. 6.4 can be written in terms of the ramp signals as

$$f(t) = 2r(t) - 2r(t-1) - 2r(t-3) + 2r(t-4)$$

where substituting $r(t) = tu(t)$, we obtain

$$f(t) = 2tu(t) - 2(t-1)u(t-1) - 2(t-3)u(t-3) + 2(t-4)u(t-4)$$

whose Laplace transform can be calculated as

$$\begin{aligned} F(s) &= 2\frac{1}{s^2} - 2\frac{e^{-s}}{s^2} - 2\frac{e^{-3s}}{s^2} + 2\frac{e^{-4s}}{s^2} \\ &= \frac{2}{s^2}\left(1 - e^{-s} - e^{-3s} + e^{-4s}\right) \end{aligned}$$

Calculation of Transfer Functions and Impulse Responses of Linear and Time-Invariant Systems by Laplace Transform

The relationship between the input and output of a linear and time-invariant continuous-time system can be expressed using differential equations. The general form of these equations is

$$\sum_{k=0}^{N} A[k]\frac{d^k y(t)}{dt^k} = \sum_{k=0}^{M} B[k]\frac{d^k x(t)}{dt^k} \tag{6.150}$$

Fig. 6.4 The graph of $f(t)$ for Example 6.28

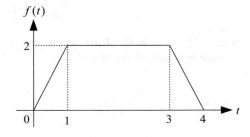

where $x(t)$ is the system input and $y(t)$ is the system output. Taking the Laplace transform of both sides of (6.150), we get

$$\left(\sum_{k=0}^{N} A[k]s^k\right) Y(s) = \left(\sum_{k=0}^{M} B[k]s^k\right) X(s) \tag{6.151}$$

from which the transfer function can be obtained as

$$H(s) = \frac{Y(s)}{X(s)} \rightarrow H(s) = \frac{\sum_{k=0}^{M} B[k]s^k}{\sum_{k=0}^{N} A[k]s^k} \tag{6.152}$$

Taking the inverse Laplace transform of the transfer function in (6.152), we obtain the impulse response of the linear and time-invariant system in the time domain, i.e., $h(t) = \mathcal{L}^{-1}\{H(s)\}$.

Causality and Stability
The roots of the denominator polynomial of the transfer function in (6.152), i.e., the roots of

$$\sum_{k=0}^{N} A[k]s^k = 0 \tag{6.153}$$

are the poles of the transfer function.

The system with transfer function $H(s)$ is causal if the region of convergence of the Laplace transform lies to the right of the rightmost pole of $H(s)$ in the complex plane. The system is stable if the region of convergence of the transfer function includes the vertical axis of the complex plane.

Hence, for a system to be both causal and stable, the real parts of all poles of the transfer function must be negative, i.e., $\text{Re}(s_i) < 0$, $i = 0 \ldots N$, and the convergence region must lie to the right of the rightmost pole covering the imaginary axis.

Example 6.29 The relationship between the input and output of a linear and time-invariant system is expressed by the differential equation

$$\frac{dy(t)}{dt} + 2y(t) = x(t) \tag{6.154}$$

where $x(t)$ is the input of the system and $y(t)$ is the output of the system. Find the impulse response of the system.

Solution 6.29 If we take the Laplace transform of both sides of (6.154), we get

$$sY(s) + 2Y(s) = X(s)$$

from which the transfer function is calculated as

$$H(s) = \frac{Y(s)}{X(s)} \rightarrow H(s) = \frac{1}{s+2}$$

Since the system is causal, the region of convergence must be to the right of the rightmost pole in the complex plane.

There is only one pole at $s = -2$, and the region of convergence can be indicated as $\text{Re}\{s\} > -2$. Considering the region of convergence, the expression of the impulse function in the time can be calculated as

$$h(t) = \mathcal{L}^{-1}\{H(s)\} \rightarrow h(t) = e^{-2t}u(t)$$

Example 6.30 The transfer function of a linear time-invariant system having causality and stability properties is given as

$$H(s) = \frac{s}{(s+3)(s+5)} \tag{6.155}$$

Determine the region of convergence for the transfer function, and find the impulse response of the system in the time domain.

Solution 6.30 The poles of the transfer function can be found from $(s+3)$ $(s+5) = 0$ as $s_1 = -3$, $s_2 = -5$. Since the system is causal and stable, the convergence region can be determined as $\text{Re}\{s\} > -3$.

The transfer function can be written as the sum of simpler polynomials as

$$\frac{s}{(s+3)(s+5)} = \frac{A}{s+3} + \frac{B}{s+5} \rightarrow A = -\frac{3}{2}, \; B = \frac{5}{2}$$

leading to

$$H(s) = -\frac{\frac{3}{2}}{s+3} + \frac{\frac{5}{2}}{s+5} \tag{6.156}$$

The impulse response of the system can be calculated using the inverse Laplace transform of (6.156). Considering the convergence region, the inverse Laplace transform of (6.156) can be calculated as

$$h(t) = \mathcal{L}^{-1}\{H(s)\} \rightarrow h(t) = -\frac{3}{2}\mathcal{L}^{-1}\left(\frac{1}{s+3}\right) + \frac{5}{2}\mathcal{L}^{-1}\left(\frac{1}{s+5}\right)$$

resulting in

$$h(t) = -\frac{3}{2}e^{-3t}u(t) + \frac{5}{2}e^{-5t}u(t)$$

Summary

The Laplace transform is used for continuous-time signals, and it is a more general form of the continuous-time Fourier transform. Some signals may not have Fourier transforms, whereas the same signals may have Laplace transforms.

The two-sided Laplace transform is calculated using

$$F(s) = \int\limits_{t=-\infty}^{\infty} f(t)e^{-st}dt \tag{6.157}$$

whereas one-sided Laplace transform is calculated using

$$F(s) = \int\limits_{t=0}^{\infty} f(t)e^{-st}dt \tag{6.158}$$

One-sided Laplace transform is used to solve differential equations. The Laplace transforms of some of the basic signals used in signal processing are shown in Table 6.1.

Let $F(s)$ and $G(s)$ be the Laplace transforms of $f(t)$ and $g(t)$. The regions of converge for $F(s)$ and $G(s)$ are denoted by ROC_f and ROC_g, respectively.

The properties of the Laplace transform are summarized in Table 6.2. These properties can be proved using the two-sided Laplace transform.

For the complex signal $f(t)$, we have

$$L\{Re\,(f(t))\} = Re\,(L\{f(t)\}) \qquad L\{Im(f(t))\} = Im(L\{f(t)\}) \tag{6.159}$$

For causal signal $f(t)$, the initial and final value theorems are stated as:
Initial value theorem:

$$f(0) = \lim_{s\to\infty} sF(s) \tag{6.160}$$

Final value theorem:

$$f(\infty) = \lim_{s\to 0} sF(s) \tag{6.161}$$

Table 6.1 Table of Laplace transforms

$f(t)$	$F(s)$	ROC
$\delta(t)$	1	s-plane
$u(t)$	$\frac{1}{s}$	$Re\{s\} > 0$
$-u(t)$	$\frac{1}{s}$	$Re\{s\} < 0$
$e^{-at}u(t)$	$\frac{1}{s+a}$	$Re\{s\} > a$
$-e^{-at}u(-t)$	$\frac{1}{s+a}$	$Re\{s\} < a$
$[\cos(at)]u(t)$	$\frac{s}{s^2+a^2}$	$Re\{s\} > 0$
$[\sin(at)]u(t)$	$\frac{a}{s^2+a^2}$	$Re\{s\} > 0$
$J_0(at)$	$\frac{1}{\sqrt{s^2+a^2}}$	$Re\{s\} > 0$
$\frac{t^n}{n!}u(t)$	$\frac{1}{s^{n+1}}$	$Re\{s\} > 0$

Table 6.2 Table of Laplace transform properties

Signal	Laplace transform	Region of convergence
$af(t) + bg(t)$	$aF(s) + bG(s)$	$ROC = ROC_f \cap ROC_g$
$f(t - t_0)$	$e^{-st_0}F(s)$	$ROC = ROC_f$
$e^{s_0 t}f(t)$	$F(s - s_0)$	$ROC = (ROC_f$ is shifted by $s_0)$
$f(at)$	$\frac{1}{|a|}F\left(\frac{s}{a}\right)$	$ROC = \frac{ROC_f}{a}$
$f(t) * g(t)$	$F(s)G(s)$	$ROC = ROC_f \cap ROC_g$
$-tf(t)$	$\frac{dF(s)}{ds}$	$ROC = ROC_f$
$\int_{-\infty}^{t} f(\tau)d\tau$	$\frac{F(s)}{s}$	$ROC = ROC_f \cap \{Re\{s\} > 0\}$

Problems

1. Find the Laplace transforms of the following signals:

   (a) $te^{2t}u(t)$ (b) $e^{-t}\cos(t)u(t)$
   (c) $\cos^2(t)u(t)$ (d) $tu^2(t - 1)$
   (e) $\delta(2t - 2)$ (f) $\sin(t) * \cos(t)$
   (g) $2\sin(2t + 1)u(t)$ (h) $t^2 e^{-t}\cos(t + 3)u(t)$
   (i) $t\cosh(t)u(t)$ (j) $t^2\sin(t)u(t)$
   (k) $e^{-t}\cos(t + 1)$

   (l) $f(t) = \begin{cases} 0 & 0 < t < \pi \\ \cos(t) & \pi \leq t \end{cases}$

   (m) $\frac{\sin(\pi t)}{\pi t}$

2. Find the inverse Laplace transforms of signals:

   (a) $\frac{s+2}{s^2+5s+6}, Re\{s\} > -2$ (b) $\frac{s+2}{(s+1)^2+4}, Re\{s\} > -1$ (c) $e^{-3s}\frac{2s+3}{s^2}$
   (d) $\frac{s}{s^2-2s+2}$ (e) $\frac{1}{2s^2+2s+1}$ (f) $\frac{1}{s^2(s^2+1)}$ (g) $\frac{1}{s^3}$ (h) $\frac{2s+1}{(s+4)^3}$

3. Prove that:
   (a) $\lim\limits_{t\to\infty} f(t) = \lim\limits_{s\to 0} sF(s)$ (b) $\lim\limits_{t\to 0} f(t) = \lim\limits_{s\to\infty} sF(s)$
4. $F(s) = \frac{s+1}{s^3 - s^2 - 2s} \to f(\infty) = ?$
5. Solve the following equations using Laplace transforms:
   (a) $f''(t) + 3f'(t) + 2f(t) = 5\delta(t - 2)$
   (b) $f(t) = e^{-4} - 4\int_0^t \cos\left(2(t - \tau)\right)f(\tau)d\tau, t > 0$
   (c) $f''(t) + f'(t) = u(t - 1), f(0) = 0, f'(0) = 15$
6. For a causal system, the input and output signals are given as $x(t) = e^{-t}u(t)$ and $y(t) = e^{-3t}\cos(t)u(t)$. Find the transfer function and impulse response of this system.
7. For $F(s) = \frac{1}{s+1} + \frac{1}{s+3}$, write the expressions for ROCs, and determine $f(t)$ for each ROC.
8. Show that for even $f(t)$, we have $F(s) = F(-s)$, and for odd $f(t)$, we have $F(-s) = -F(s)$.
9. The pole-zero diagram of a transfer function is shown in Fig. P6.1. Identify all possible convergence regions, and for each convergence region, determine whether the system is causal and stable or not.

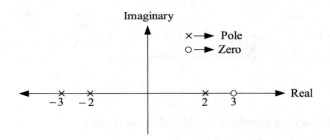

Fig. P6.1 The graph of the convergence region for Problem 9

10. If $g(t) = f(t) * f(-t)$, then express $G(s)$ in terms of $F(s)$.
11. The Laplace transform of the right-sided $f(t)$ is given as $F(s) = \frac{s}{s^2+1}$. Find the Laplace transforms of the following signals:
    (a) $f(3t)$ (b) $e^{-2t}f(t)$ (c) $tf(t)$ (d) $f(2t - 3)$ (e) $\int_0^t f(\tau)d\tau$
12. The convergence region of the transfer function of a linear and time-invariant system is shown in Fig. P6.2. Find the transfer function and impulse response of the system. Determine whether the system is causal and stable.

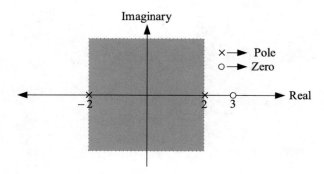

Fig. P6.2 The graph of the convergence region for Problem 12

13. The differential equations showing the relationships between inputs and outputs of some linear time-invariant and causal systems are given as:

(a) $\frac{dy(t)}{dt} + 5y(t) = 5x(t)$ (b) $\frac{d^2y(t)}{dt^2} + 7y(t) + 6y(t) = x(t) + \frac{dx(t)}{dt}$

Find the transfer function $H(s)$ and the impulse response $h(t)$ of each system.

Chapter 7
Z Transform

In the previous chapter, we explained the Laplace transform. The Laplace transform was a more general form of the Fourier transform of continuous-time signals. In a similar manner, the Z transform can be considered as a more general form of the Fourier transform used for digital signals. Discrete-time Fourier transforms of impulse responses of linear time-invariant and unstable systems cannot be calculated, whereas Z transforms of impulse responses of the same systems can be calculated. That is, the set of signals to which the Z transform can be applied includes the set of signals to which the Fourier transform can be applied. In other words, the Fourier transform is a subset of the Z transform. We cannot say that Z transform necessarily exists for every signal. For some signals, the Z transform converges for certain Z values. The region where the Z transform converges in the complex plane is called the region of convergence. The Z transforms of two different signals can be the same polynomial; however, the convergence regions may be different. In the block diagram representation of digital communication systems, the Z transform expression is generally used. The Z transform of the input signal, the transfer function of the system, and the Z transform of the output of the system are used to explain the operation of the system.

In the next section, we provide the definition of Z transform and the relationship between Z transform and discrete-time Fourier transform. In succeeding sections, we explain finding Z transform of some special signals and determination of their convergence regions. Determining the properties of linear and time-invariant systems from the Z transform of their impulse functions is covered next. Inverse Z transform is another topic covered in this chapter. Inverse Z transform is used to find the function in time domain using its Z transform. Although the inverse Z transform can be calculated using inverse Z transform formula, it is not a commonly used formula. Instead, by using the known Z transforms by deliberation, the time domain signal for which the inverse Z transform is desired is found. We also explain the properties of the Z transform. All of these properties can be proved using the definition of Z transform. Discrete-time linear and time-invariant systems can be

© The Author(s), under exclusive license to Springer Nature Switzerland AG 2023   303
O. Gazi, *Principles of Signals and Systems*, https://doi.org/10.1007/978-3-031-17789-7_7

expressed with difference equations. The difference equations of such systems can be solved using their Z transforms, and in sequel the impulse responses of these systems can be determined. We will address this issue in the last part of the chapter.

7.1 *Z* Transform

The Z transform of the digital signal $f[n]$ is calculated using

$$F(z) = \sum_{n=-\infty}^{\infty} f[n]z^{-n} \tag{7.1}$$

or using

$$F(z) = \frac{1}{\sqrt{2\pi}} \sum_{n=-\infty}^{\infty} f[n]z^{-n} \tag{7.2}$$

where z is a complex variable. Z transform can be expressed using the pair

$$f[n] \overset{Z}{\leftrightarrow} F(z) \tag{7.3}$$

In this book, we will use the definition in (7.1) for the calculation of the Z transform. z complex variable can be written as

$$z = |z|e^{j\Omega} \tag{7.4}$$

z complex variable can be indicated in the complex plane by a circle with a radius $|z|$ as shown in Fig. 7.1.

Fig. 7.1 z plane

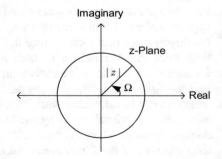

If we substitute $z = |z|e^{j\Omega}$ in (7.1), we get

$$
\begin{aligned}
F(z) &= \sum_{n=-\infty}^{\infty} f[n](|z|e^{j\Omega})^{-n} \\
&= \sum_{n=-\infty}^{\infty} (f[n]|z|^{-n})e^{-j\Omega n}
\end{aligned}
\tag{7.5}
$$

where is seen that the Z transform of $f[n]$ is nothing but the Fourier transform of $f[n]|z|^{-n}$. For $|z| = 1$, the Z transform equals the Fourier transform. Then the discrete-time Fourier transform, i.e., digital Fourier transform, is a special case of Z transform calculated on the unit circle. Hence, the Z transform can be expressed as

$$
F(z) = \mathrm{FT}\{f[n]|z|^{-n}\}
\tag{7.6}
$$

where FT indicates the Fourier transform and the relationship between Fourier and Z transforms can be expressed using

$$
F(\Omega) = (F(z)|_{z=e^{j\Omega}}
\tag{7.7}
$$

The Z transform of $f[n]$

$$
F(z) = \sum_{n=-\infty}^{\infty} f[n]z^{-n}
\tag{7.8}
$$

may not have a finite value, or it can be finite for a set of z numbers falling into an interval. Determining the range of z values in which the Z transform of the $f[n]$ takes finite values is called finding the convergence region of the Z transform. The region of convergence, ROC, of the Z transform is indicated by a circle in the complex plane. If the convergence region includes the unit circle, then Fourier transform of the signal also exists.

If the unit circle does not lie within the region of convergence, then although the Z transform is available, the Fourier transform does not exist, i.e., it is divergent.

Example 7.1 Find the Z transform of $f[n] = \alpha^n u[n]$ where $u[n]$ is the unit step signal.

Solution 7.1 The Z transform of the given signal can be calculated as

$$
F(z) = \sum_{n=-\infty}^{\infty} \alpha^n u[n]z^{-n}
\tag{7.9}
$$

which can be written as

$$
F(z) = \sum_{n=0}^{\infty} \alpha^n z^{-n}
\tag{7.10}
$$

leading to

$$F(z) = \sum_{n=0}^{\infty} \left(\alpha z^{-1} \right)^n \tag{7.11}$$

where the summation expression converges if we have

$$\left| \alpha z^{-1} \right| < 1 \rightarrow |z| > \alpha \tag{7.12}$$

For $|z| > \alpha$, using the calculus property

$$\sum_{n=0}^{\infty} r^n = \frac{1}{1-r}, |r| < 1 \tag{7.13}$$

(7.11) can be calculated as

$$F(z) = \frac{1}{1 - \alpha z^{-1}} \tag{7.14}$$

The convergence region is indicated by $|z| > \alpha$ which shows the outer region of the circle with radius α. In Fig. 7.2, the convergence region for $\alpha = 0.5$ is shown.

It is seen from Fig. 7.2 that the convergence region for $\alpha = 0.5$ encloses the unit circle and then the Fourier transform of $f[n] = 0.5^n u[n]$ also converges.

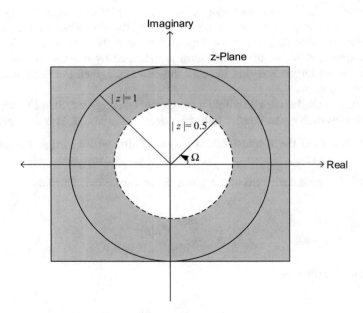

Fig. 7.2 The convergence region for the Z transform of $f[n] = 0.5^n u[n]$

Example 7.2 $f[n]$ is given as

$$f[n] = -\alpha^n u[-n-1] \tag{7.15}$$

Find $F(z)$.

Solution 7.2 Z transform of $f[n]$ can be calculated as

$$F(z) = \sum_{n=-\infty}^{\infty} -\alpha^n u[-n-1]z^{-n}$$

leading to

$$F(z) = -\sum_{n=-\infty}^{-1} \alpha^n z^{-n}$$

where changing the limits of the summation, we get

$$F(z) = -\sum_{n=1}^{\infty} \alpha^{-n} z^n$$

leading to

$$F(z) = 1 - \sum_{n=0}^{\infty} \left(\alpha^{-1} z\right)^n$$

where the summation converges for

$$\left|\alpha^{-1} z\right| < 1 \rightarrow |z| < \alpha$$

which indicates the convergence region and $F(z)$ is calculated as

$$F(z) = 1 - \sum_{n=0}^{\infty} \left(\alpha^{-1} z\right)^n \rightarrow F(z) = 1 - \frac{1}{1-\alpha^{-1} z} \rightarrow F(z) = \frac{z}{z-\alpha}$$

For $\alpha = 0.5$, the convergence region is depicted in Fig. 7.3.

Since the convergence region for $\alpha = 0.5$ does not enclose the unit circle in Fig. 7.3, the Fourier transform of $f[n] = 0.5^n u[n]$ does not exist, i.e., it diverges.

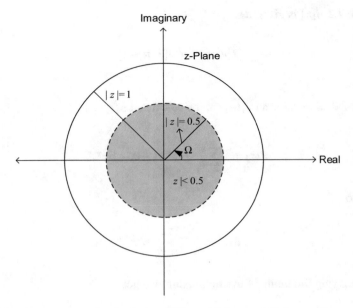

Fig. 7.3 The convergence region for the Z transform of $f[n] = -0.5^n u[-n-1]$

Example 7.3 $f[n] = \left(\frac{1}{2}\right)^n u[n] + \left(\frac{1}{4}\right)u[n] \rightarrow F(z) = ?$

Solution 7.3 Using the result

$$a^n u[n] \overset{Z}{\leftrightarrow} \frac{1}{1-az^{-1}}, \quad |z| > a$$

for $f[n]$, we get

$$\left(\frac{1}{2}\right)^n u[n] \overset{Z}{\leftrightarrow} \frac{1}{1-\frac{1}{2}z^{-1}}, \quad |z| > \frac{1}{2} \qquad \left(\frac{1}{4}\right)^n u[n] \overset{Z}{\leftrightarrow} \frac{1}{1-\frac{1}{4}z^{-1}}, \quad |z| > \frac{1}{4} \quad (7.16)$$

and accordingly $F(z)$ can be written as

$$F(z) = \frac{1}{1-\frac{1}{2}z^{-1}} + \frac{1}{1-\frac{1}{4}z^{-1}}, \quad |z| > \frac{1}{2} \qquad (7.17)$$

where the region of convergence is found by taking the intersection of the region of convergences in (7.16) as

$$\left(|z| > \frac{1}{2}\right) \cap \left(|z| > \frac{1}{4}\right) \rightarrow |z| > \frac{1}{2} \qquad (7.18)$$

7.1.1 Properties of the Convergence Region

The convergence region of the Z transform has certain properties, and these properties can be listed as follows:

1. The region of convergence has an annular shape in the z plane.
2. There is no pole of the Z transform polynomial in the region of convergence.
3. If $f[n]$ is a finite-length digital signal, the region of convergence is the entire z plane or the entire z plane excluding $z = 0$ and $z = \infty$.
4. For right-sided digital signals, the convergence regions of the Z transforms are denoted either by $|z| > z_0$ or by $|z| \leq z_0$.
5. For left-sided digital signals, the convergence regions of the Z transforms are denoted either by $|z| < z_0$ or by $|z| \leq z_0$.
6. For two-sided digital signals, the convergence regions of the Z transforms are denoted either by $0 < |z| < z_0$ or by $0 < |z| \leq z_0$.

Example 7.4 Find the z transform and determine the region of convergence of

$$f[n] = \begin{cases} \alpha^n & 0 \leq n < N, \ \alpha > 0 \\ 0 & \text{otherwise} \end{cases} \tag{7.19}$$

Solution 7.4 We can employ the Z transform formula for the given signal as in

$$F(z) = \sum_{m=-\infty}^{\infty} f[n]z^{-n}$$
$$= \sum_{n=0}^{N-1} \alpha^n z^{-n}$$

leading to

$$F(z) = \sum_{n=0}^{N-1} (\alpha z^{-1})^n$$

where employing the property

$$\sum_{n=0}^{N-1} r^n = \frac{1 - r^N}{1 - r}, \quad |r| < 1$$

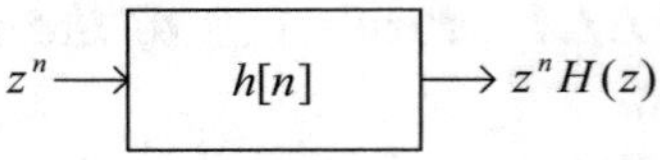

Fig. 7.4 A LTI system

we get

$$F(z) = \frac{1 - (\alpha z^{-1})^N}{1 - \alpha z^{-1}}$$

which can be rearranged as

$$F(z) = \frac{1}{z^{N-1}} \frac{z^N - \alpha^N}{z - \alpha} \tag{7.20}$$

When (7.20) is inspected, it is seen that $F(z)$ has poles as $z = 0$ and $z = a$ and the convergence region is the entire z plane excluding the points $z = 0$ and $z = \alpha$.

Example 7.5 Let $h[n]$ be the impulse response of a linear time-invariant system. Find the output of this system for the input $x[n] = z^n$ where z is a complex variable.

Solution 7.5 The output of a linear time-invariant system with impulse response $h[n]$ for an arbitrary input $x[n]$ can be calculated as

$$y[n] = \sum_{k=-\infty}^{\infty} h[k]x[n-k]$$

which can be calculated for the input signal $x[n] = z^n$ as

$$\begin{aligned} y[n] &= \sum_{k=-\infty}^{\infty} h[k]z^{n-k} \\ &= z^n \sum_{k=-\infty}^{\infty} h[k]z^{-k} \\ &= z^n H(z) \end{aligned}$$

Thus, for a linear and time-invariant system with impulse response $h[n]$, for input $x[n] = z^n$, the system output is $y[n] = z^n H(z)$ where $H(z)$ is the Z transform of $h[n]$. This property is illustrated in Fig. 7.4.

7.2   Inverse Z Transform

The inverse Z transform is used to find the digital sequence $f[n]$ using its Z transform $F(z)$. To find $f[n]$ a number of methods can be considered. In this section, we explain how to find the inverse Z transform using different methods.

Inverse Z Transform Formula

The Z transform of $f[n]$ is calculated using

$$F(z) = \sum_{n=-\infty}^{\infty} f[n]z^{-n} \tag{7.21}$$

where $z = |z|e^{j\Omega}$, and the relationship between Z transform and Fourier transform can be written as

$$F(z) = \text{FT}\{f[n]|z|^{-n}\} \tag{7.22}$$

from which we get

$$f[n]|z|^{-n} = \text{FT}^{-1}\{F(z)\} \tag{7.23}$$

where using the definition of inverse Fourier transform

$$g[n] = \frac{1}{2\pi} \int_{2\pi} \widehat{g}(\Omega)e^{j\Omega n} d\Omega$$

which corresponds to the Z transform definition in (7.21), we obtain

$$f[n]|z|^{-n} = \frac{1}{2\pi} \int_{2\pi} F(z)e^{j\Omega n} d\Omega \tag{7.24}$$

leading to

$$f[n] = \frac{1}{2\pi} \int_{2\pi} F(z)\left(|z|e^{j\Omega}\right)^n d\Omega \tag{7.25}$$

where the differential term $d\Omega$ can be expressed in terms of z parameter as

$$z = |z|e^{j\Omega} \rightarrow dz = j|z|e^{j\Omega} d\Omega \rightarrow dz = jz d\Omega \rightarrow d\Omega = \frac{1}{j}z^{-1}dz \tag{7.26}$$

and substituting $d\Omega = \frac{1}{j}z^{-1}dz$ in (7.25), we obtain

$$f[n] = \frac{1}{j2\pi} \int_{2\pi} F(z)z^{n-1} dz \tag{7.27}$$

where the integral is calculated in the complex plane along a circle and different circles with different radiuses can be considered. Considering circles with different radiuses, equation (7.27) can be written as

$$f[n] = \frac{1}{j2\pi} \oint F(z)z^{n-1}dz \tag{7.28}$$

where the contour integral is evaluated along a circle centered at origin and this circle includes the convergence region.

The contour integration

$$f[n] = \frac{1}{j2\pi} \oint F(z)z^{n-1}dz \tag{7.29}$$

can be evaluated using the Cauchy residue theorem as in

$$
\begin{aligned}
f[n] &= \frac{1}{i2\pi} \oint F(z)z^{n-1}dz \\
&= \left(\sum \text{pole residues of } F(z)z^{n-1} \text{ inside the contour} \right)
\end{aligned}
\tag{7.30}
$$

which can be written in a simple form as

$$f[n] = \sum \text{Re} \, s\left(z^{n-1}F(z)\right) \tag{7.31}$$

Although (7.31) is emphasized by many authors, it is not used much because its application is not practical and difficult to calculate for some Z transforms.

Example 7.6 Show that the result of contour integral

$$\frac{1}{j2\pi} \oint F(z)z^{n-1}dz \tag{7.32}$$

equals $f[n]$.

Solution 7.6 Substituting

$$F(z) = \sum_{k=-\infty}^{\infty} f[k]z^{-k}$$

in

$$\frac{1}{j2\pi} \oint F(z)z^{n-1}dz$$

we get

$$\frac{1}{j2\pi} \oint \sum_{k=-\infty}^{\infty} f[k]z^{-k}z^{n-1}dz$$

which can be written as

$$\sum_{k=-\infty}^{\infty} f[k]\frac{1}{j2\pi} \oint z^{-k+n-1}dz$$

where using the Cauchy integral theorem,

$$\oint_C z^{-k}dz = j2\pi\delta[k-1]$$

we obtain

$$\sum_{k=-\infty}^{\infty} f[k]\frac{1}{j2\pi} \oint z^{-k+n-1}dz \rightarrow \sum_{k=-\infty}^{\infty} f[k]\delta[n-k]$$

resulting in $f[n]$.

Example 7.7 Find the pole residues of

$$z^{n-1}F(z) \tag{7.33}$$

where

$$F(z) = \frac{4z}{3z^2 - 2z - 1} \tag{7.34}$$

Solution 7.7 The denominator of the polynomial

$$F(z) = \frac{4z}{3z^2 - 2z - 1}$$

can be factorized as

$$F(z) = \frac{4z}{(3z+1)(z-1)}$$

then we can write

$$z^{n-1}F(z) = \frac{4z^n}{(3z+1)(z-1)}$$

whose poles are $z = 1$ and $z = -\frac{1}{3}$. Pole residues can be calculated as

$$\mathrm{Re}\,s_{z=1}\left(z^{n-1}F(z)\right) = \left((z-1)z^{n-1}F(z)\right)\big|_{z=1}$$

$$= \frac{4z^n}{3z+1}\bigg|_{z=1}$$

$$= 1$$

$$\mathrm{Re}\,s_{z=-\frac{1}{3}}\left(z^{n-1}F(z)\right) = \left((z+1)z^{n-1}F(z)\right)\big|_{z=1}$$

$$= \frac{1}{3}\frac{4z^n}{(z-1)}\bigg|_{z=-\frac{1}{3}}$$

$$= -\left(-\frac{1}{3}\right)^n$$

Example 7.8 Z transform of a digital signal is given as

$$F(z) = \frac{4z}{3z^2 - 2z - 1}, \quad |z| > \frac{1}{3} \tag{7.35}$$

Find $f[n]$.

Solution 7.8 $F(z)$ can be written as

$$F(z) = \frac{4z}{(3z+1)(z-1)}$$

where it is seen that the poles of $F(z)$ are $z = 1$ and $z = -\frac{1}{3}$. Using the formula

$$f[n] = \sum \mathrm{Re}\,s\left(z^{n-1}F(z)\right)$$

considering the convergence region $|z| > \frac{1}{3}$, we get

$$f[n] = \left(\mathrm{Re}\,s_{z=1}\left(z^{n-1}F(z)\right) + \mathrm{Re}\,s_{z=-\frac{1}{3}}\left(z^{n-1}F(z)\right)\right)$$

resulting in

$$f[n] = \left(1 - \left(-\frac{1}{3}\right)^n\right)u[n]$$

Using Known Z Transform Pairs
It is a commonly used method to find the digital signal in the time domain from
its Z transform. In this method, the signal in the time domain is calculated using
the known Z transforms.

Example 7.9 Find the inverse Z transform of

$$F(z) = \frac{1}{\left(1 - \frac{1}{2}z^{-1}\right)\left(1 - 3z^{-1}\right)}, \quad |z| > 3 \tag{7.36}$$

Solution 7.9 The rational polynomial $F(z)$ can be written as

$$\frac{1}{\left(1 - \frac{1}{2}z^{-1}\right)\left(1 - 3z^{-1}\right)} = \frac{A}{1 - \frac{1}{2}z^{-1}} + \frac{B}{1 - 3z^{-1}}$$

where the coefficients can be calculated as $A = -\frac{1}{5}$, $B = \frac{6}{5}$, and accordingly we get

$$F(z) = -\frac{1}{5}\frac{1}{1 - \frac{1}{2}z^{-1}} + \frac{6}{5}\frac{1}{1 - 3z^{-1}} \tag{7.37}$$

whose inverse Z transform can be calculated using the property

$$\alpha^n u[n] \overset{Z}{\leftrightarrow} \frac{1}{1 - \alpha z^{-1}}, \quad |z| > \alpha \tag{7.38}$$

as

$$f[n] = -\frac{1}{5}\left(\frac{1}{2}\right)^n u[n] + \frac{6}{5}(3)^n u[n] \tag{7.39}$$

Note that while calculating the inverse Z transform of (7.37), we have to pay
attention to the convergence region. It is known that

$$-\alpha^n u[-n-1] \overset{Z}{\leftrightarrow} \frac{1}{1 - \alpha z^{-1}}, \quad |z| < \alpha \tag{7.40}$$

and when the convergence region of $F(z)$ is inspected, we see that the pair

$$-\alpha^n u[-n-1] \overset{Z}{\leftrightarrow} \frac{1}{1 - \alpha z^{-1}}, \quad |z| < \alpha \tag{7.41}$$

is not useful for the calculation of inverse Z transform and we use the pair

$$a^n u[n] \overset{Z}{\leftrightarrow} \frac{1}{1 - az^{-1}}, \quad |z| > a \tag{7.42}$$

for the calculation of inverse Z transform.

Example 7.10 Calculate the inverse Z transform of

$$F(z) = \ln\left(1 + az^{-1}\right), \quad |z| > |a| \tag{7.43}$$

Solution 7.10 To solve this problem, we will use the Taylor expansion of the logarithmic signal. Taylor expansion of the logarithmic signal can be achieved using

$$\ln(1 + z) = \sum_{n=1}^{\infty} \frac{(-1)^{n+1} z^n}{n}, |z| < 1 \tag{7.44}$$

Using (7.44) in (7.43), we obtain

$$F(z) = \sum_{n=1}^{\infty} \frac{(-1)^{n+1} a^n z^{-n}}{n} \tag{7.45}$$

when compared to

$$F(z) = \sum_{n=-\infty}^{\infty} f[n] z^{-n} \tag{7.46}$$

we get

$$f[n] = \begin{cases} (-1)^{n+1} \dfrac{a^n}{n} & n \leq 1 \\ 0 & \text{otherwise} \end{cases} \tag{7.47}$$

which can be rearranged as

$$f[n] = \begin{cases} -\dfrac{(-a)^n}{n} & n \leq 1 \\ 0 & \text{otherwise} \end{cases} \tag{7.48}$$

Example 7.11 Find the Z transforms of $f[n] = \delta[n]$ and $g[n] = \delta[n - n_0]$.

Solution 7.11 Employing Z transform formula for $f[n]$, we get

$$
\begin{aligned}
F(z) &= \sum_{n=-\infty}^{\infty} f[n]z^{-n} \\
&= \sum_{n=-\infty}^{\infty} \delta[n]z^{-n}
\end{aligned}
$$

where using the property

$$
\sum_{n=-\infty}^{\infty} f[n]\delta[n - n_0] = f[n_0]
$$

we obtain

$$
F(z) = z^0 \rightarrow F(z) = 1
$$

In a similar manner, the Z transform of $g[n]$ can be calculated as

$$
\begin{aligned}
G(z) &= \sum_{n=-\infty}^{\infty} g[n]z^{-n} \\
&= \sum_{n=-\infty}^{\infty} \delta[n - n_0]z^{-n} \\
&= z^{-n_0}
\end{aligned}
$$

which can be written with a transform pair as

$$
\delta[n - n_0] \overset{Z}{\leftrightarrow} z^{-n_0} \tag{7.49}
$$

Polynomial Division or Polynomial Expansion
In this method, the digital signal is determined by comparing the definition of Z transform formula to the polynomial expansion of the Z transform of the digital signal. If the Z transform expression is a rational polynomial, division is performed and written as the sum of the powers of z.

Example 7.12 $X(z) = 3z^3 - 2 + 4z^{-2}, 0 < |z| < \infty \rightarrow x[n] = ?$

Solution 7.12 If we expand the summation in Z transform formula

$$
X(z) = \sum_{n=-\infty}^{\infty} x[n]z^{-n}
$$

and compare it to the $X(z) = 3z^3 - 2 + 4z^{-2}$, we find that

$$x[n] = \begin{cases} 3 & n = -3 \\ -2 & n = 0 \\ 4 & n = 2 \\ 0 & \text{otherwise} \end{cases}$$

which can be written in terms of impulse signals as

$$x[n] = 3\delta[n+3] - 2\delta[n] + 4\delta[n-2]$$

Example 7.13 $X(z) = \frac{1}{1-z^{-1}}, \quad |z| < 1 \rightarrow x[n] = ?$

Solution 7.13 The polynomial division for $1/(1 - z^{-1})$ can be performed as

$$-z^{-1}+1\overline{)\begin{array}{l} \quad\quad -z - z^2 - \cdots \\ \quad 1 \\ \quad \underline{1-z} \\ \quad\quad z \end{array}}$$

resulting in

$$\frac{1}{1-z^{-1}} = -z - z^2 - \cdots \tag{7.50}$$

When (7.50) is compared to

$$X(z) = \sum_{n=-\infty}^{\infty} x[n]z^{-n} \tag{7.51}$$

we find that $x[n] = -u[-n-1]$.

Note The polynomial division can also be achieved as

$$\frac{1}{1-z^{-1}} = 1 + z^{-1} + z^{-2} + \cdots \tag{7.52}$$

However, such an expansion does not converge for $|z| < 1$.

Exercise $X(z)$ is given as

$$X(z) = \frac{1}{1-\alpha z^{-1}}, \quad |z| > \alpha \tag{7.53}$$

Find $x[n]$.

Properties of Z Transform

Let $F(z)$ and $G(z)$ be the Z transforms of $f[n]$ and $g[n]$ with convergence regions ROC_f and ROC_g. Using these definitions, we can list the following properties of the Z transform.

1. Linearity:

$$\alpha f[n] + \beta g[n] \overset{Z}{\leftrightarrow} \alpha F(z) + \beta G(z), \quad \mathrm{ROC} = \mathrm{ROC}_f \cap \mathrm{ROC}_g \qquad (7.54)$$

2. Shifting in the time domain:

$$f[n - n_0] \overset{Z}{\leftrightarrow} z^{-n_0} F(z), \quad \mathrm{ROC} = \mathrm{ROC}_f \qquad (7.55)$$

   where the new convergence region may not include origin or infinity.
3. Shifting in frequency:

$$z_0^n f[n] \overset{Z}{\leftrightarrow} F\left(\frac{z}{z_0}\right), \quad z_0 = e^{j\Omega_0}, \quad \mathrm{ROC} = z_0 \mathrm{ROC}_f \qquad (7.56)$$

4. Time reversal:

$$f[-n] \overset{Z}{\leftrightarrow} F\left(\frac{1}{z}\right), \quad \mathrm{ROC} = \frac{1}{\mathrm{ROC}_f} \qquad (7.57)$$

5. Differentiation in Z domain:

$$nf[n] \overset{Z}{\leftrightarrow} -z\frac{dF(z)}{dz}, \quad \mathrm{ROC} = \mathrm{ROC}_f \qquad (7.58)$$

6. Z Transform of the conjugate signal:

$$f^*[n] \overset{Z}{\leftrightarrow} F^*(z), \quad \mathrm{ROC} = \mathrm{ROC}_f, \quad f^*[-n] \overset{Z}{\leftrightarrow} F^*\left(\frac{1}{z}\right), \quad \mathrm{ROC} = \frac{1}{\mathrm{ROC}_f} \qquad (7.59)$$

7. *Z* transform of convolution:

$$f[n] * g[n] \overset{Z}{\leftrightarrow} F(z)G(z), \quad \text{ROC} = \text{ROC}_f \cap \text{ROC}_g \tag{7.60}$$

8. Convolution property in *Z* domain:

$$f[n]g[n] \overset{Z}{\leftrightarrow} \frac{1}{2\pi j} \oint_C F(v)G\left(\frac{z}{v}\right)v^{-1}dv \tag{7.61}$$

where *C* is a contour on the intersection of convergence regions of $F(z)$ and $G(z)$ around the origin.

Initial Value Theorem
For causal $f[n]$ signal, we have

$$f[0] = \lim_{z \to \infty} F(z) \tag{7.62}$$

Proof The *Z* transform

$$F(z) = \sum_{n=0}^{\infty} f[n]z^{-n}$$

of the causal $f[n]$, i.e., $f[n] = 0$, $n < 0$, $F(z)$ can be written as

$$F(z) = f[0] + f[1]z^{-1} + f[2]z^{-2} + f[3]z^{-3} + \ldots$$

where it is seen that

$$f[0] = \lim_{z \to \infty} F(z) \tag{7.63}$$

Final Value Theorem
The final value theorem is used to find the last value of the signal, and it is stated as

$$\lim_{n \to \infty} f[n] = \lim_{z \to 1} (z-1)F(z) \tag{7.64}$$

7.3   Z Transform of Linear and Time-Invariant Digital Systems

The relationship between the output $y[n]$ and the input $x[n]$ of a digital system is expressed as $y[n] = x[n] * h[n]$ where $h[n]$ is the impulse response of the system.

The Z transform of the output of the system can be written as

$$Y(z) = X(z)H(z) \qquad (7.65)$$

from which we get

$$H(z) = \frac{Y(z)}{X(z)} \qquad (7.66)$$

where $H(z)$ is the Z transform of $h[n]$ and it is called the transfer function of the system. We can determine the other properties of a linear and time-invariant system from its transfer function.

Causality Property
The criteria for a linear and time-invariant system to be causal is

$$h[n] = 0, n < 0 \qquad (7.67)$$

If Z transform of $h[n]$

$$H(z) = \sum_{n=0}^{\infty} h[n]z^{-n} \qquad (7.68)$$

is inspected considering the constraint (7.67), it is seen that $H(z)$ contains only negative powers of z, and the convergence region of $H(z)$ extends to infinity.

We conclude that if the region of convergence of the transfer function of the system is a region extending from the boundary of a circle toward infinity, then the system has causal property.

Stability Property
The criteria for a linear and time-invariant system to be stable is

$$\sum |h[n]| < \infty \qquad (7.69)$$

If the impulse response of the system is absolutely summable, then the Fourier transform of the impulse response of the system can be calculated. If the impulse response of the system has a Fourier transform, this means that the convergence region of the Z transform includes the unit circle. Hence, we can conclude that for a

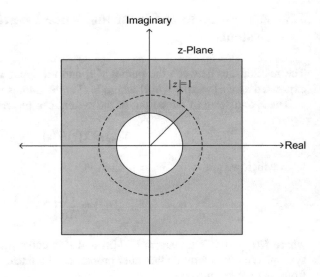

Fig. 7.5 Convergence region of the Z transform of the impulse response of a linear, time-invariant, stable, and causal system

linear and time-invariant system to be a stable system, the convergence region of the Z transform of the impulse response must include the unit circle.

The condition for a linear and time-invariant system to be causal and stable is that the convergence region of the Z transform of the impulse response must include the unit circle and it must cover a region extending to infinity. The convergence region of a linear and time-invariant system having causality and stability properties is depicted in Fig. 7.5.

Finding the Impulse Responses of Linear and Time-Invariant Systems from Difference Equations

The relationship between inputs and outputs of the linear and time-invariant systems can be described by the difference equations

$$\sum_{k=0}^{N} a_k y[n-k] = \sum_{k=0}^{M} b_k x[n-k] \tag{7.70}$$

which can be used to calculate the transfer functions of the systems.

Evaluating the Z transform of both sides of (7.70), we get

$$Y(z) \sum_{k=0}^{N} a_k z^{-k} = X(z) \sum_{k=0}^{M} b_k z^{-k} \tag{7.71}$$

from which we obtain

$$H(z) = \frac{Y(z)}{X(z)} \rightarrow H(z) = \frac{\sum_{k=0}^{M} b_k z^{-k}}{\sum_{k=0}^{N} a_k z^{-k}}. \tag{7.72}$$

Taking the inverse Z transform of the rational expression in (7.72), we get the impulse response of the system in the time domain. $H(z)$ in (7.72) can be written as the product of simple polynomials as explained below:

1. $H(z)$ can be written as

$$H(z) = \frac{b_0 \prod_{k=1}^{M} (1 - c_k z^{-1})}{a_0 \prod_{k=1}^{N} (1 - d_k z^{-1})} \tag{7.73}$$

where c_k and d_k are the zeros and poles of the rational polynomial.
2. If $M < N$, then $H(z)$ can be written as

$$H(z) = \sum_{k=1}^{N} \frac{A_k}{1 - d_k z^{-1}} \tag{7.74}$$

where the coefficients A_k are calculated using

$$A_k = \left(1 - d_k z^{-1}\right) H(z) \big|_{z = d_k} \tag{7.75}$$

3. If $M \leq N$, then $H(z)$ can be written as

$$H(z) = \sum_{l=0}^{M-N} B_l z^{-1} + \sum_{k=1}^{N} \frac{A_k}{1 - d_k z^{-1}} \tag{7.76}$$

where the coefficients B_l are obtained from the polynomial division and the coefficients A_k are calculated as for $M < N$ case explained in (2).
4. The general expression of the partial fraction expansion operation, considering multiple poles, can be written as

$$H(z) = \sum_{l=0}^{M-N} B_l z^{-l} + \sum_{k=1}^{N-S} \frac{A_k}{1 - d_k z^{-1}} + \sum_{m=1}^{S} \frac{C_m}{(1 - r z^{-1})m} \tag{7.77}$$

where m is used for the degrees of multiple poles and the coefficients C_m are calcualted as

$$C_m = \frac{1}{(S-m)!} \left(-\frac{1}{r}\right)^{S-m} \frac{d^{S-m}H(z)}{dz^{S-m}} \left((1-rz^{-1})^S H(z)\right)\Big|_{z=r^{-1}} \qquad (7.78)$$

Example 7.14 Find the inverse Z transform of

$$F(z) = \frac{1 + 2z^{-1} + z^{-2}}{1 - \frac{3}{2}z^{-1} + \frac{1}{2}z^{-2}}, \quad |z| > 1 \qquad (7.79)$$

Solution 7.14 $F(z)$ in (7.79) can be written as

$$F(z) = 2 + \frac{-1 + 5z^{-1}}{1 - \frac{3}{2}z^{-1} + \frac{1}{2}z^{-2}}$$

where the denominator can be expressed as the product of polynomials as

$$F(z) = 2 + \frac{-1 + 5z^{-1}}{\left(1 - \frac{1}{2}z^{-1}\right)\left(1 - z^{-1}\right)}$$

which can be written as

$$F(z) = 2 + \frac{A_1}{1 - \frac{1}{2}z^{-1}} + \frac{A_2}{1 - z^{-1}}$$

where the coefficients A_1 and A_2 can be calculated as

$$A_1 = \left(F(z)\left(1 - \frac{1}{2}z^{-1}\right)\right)\Big|_{z^{-1}=2} = \frac{1 + 4 + 4}{1 - 2} = -9$$

$$A_2 = \left(F(z)\left(1 - z^{-1}\right)\right)\Big|_{z^{-1}=1} = \frac{1 + 2 + 1}{\frac{1}{2}} = 8$$

Hence, we obtain the partial fraction expansion of $F(z)$ as

$$F(z) = 2 - \frac{9}{1 - \frac{1}{2}z^{-1}} + \frac{8}{1 - z^{-1}} \qquad (7.80)$$

The inverse Z transform of (7.80) considering the convergence region $|z| > 1$ can be calculated as

$$f[n] = 2\delta[n] - 9\left(\frac{1}{2}\right)^n u[n] + 8u[n] \tag{7.81}$$

Example 7.15 $F(z) = \ln(1 + \alpha z^{-1})$, $|z| > |\alpha|$, find $f[n]$.

Solution 7.15 We solved this question in our previous examples. We will solve the same question in a different way using the derivative property of the Z transform.

We have the transform pair

$$nf[n] \overset{Z}{\leftrightarrow} -z\frac{dF(z)}{dz} \tag{7.82}$$

where using $F(z) = \ln(1 + \alpha z^{-1})$, $|z| > |\alpha|$, we obtain

$$nf[n] \overset{Z}{\leftrightarrow} -z\frac{d(\ln(1 + \alpha z^{-1}))}{dz} \rightarrow \frac{\alpha z^{-1}}{a + \alpha z^{-1}}, \quad |z| > |\alpha| \tag{7.83}$$

We have the transform pair

$$\alpha^n u[n] \overset{Z}{\leftrightarrow} \frac{1}{1 - \alpha z^{-1}}, \quad |z| > |\alpha| \tag{7.84}$$

and

$$f[n - n_0] \overset{Z}{\leftrightarrow} z^{-n_0} F(z) \tag{7.85}$$

Accordingly, the signal with Z transform

$$\frac{\alpha z^{-1}}{1 + \alpha z^{-1}}, \quad |z| > |\alpha| \tag{7.86}$$

can be determined using (7.84) and (7.85) as

$$\alpha(-\alpha)^{n-1} u[n-1] \tag{7.87}$$

which is equal to $nf[n]$. Thus, we obtain

$$\begin{aligned}
f[n] &= \frac{1}{n}\alpha(-\alpha)^{n-1} u[n-1] \\
&= -\frac{1}{n}(-\alpha)(-\alpha)^{n-1} u[n-1] \\
&= -\frac{(-\alpha)^n}{n} u[n-1]
\end{aligned}$$

Example 7.16 $f[n] = \alpha^n \cos(\beta n) u[n] \rightarrow F(z) = ?$

Solution 7.16 Using

$$\cos(\beta n) = \frac{1}{2}\left(e^{j\beta n} + e^{-j\beta n}\right) \tag{7.88}$$

in $f[n] = \alpha^n \cos(\beta n) u[n]$, we get

$$f[n] = \frac{1}{2}\left[\left(\alpha e^{j\beta}\right)^n + \left(\alpha e^{-j\beta}\right)^n\right]u[n]$$

The Z transform of $f[n]$ can be calculated using the transform pairs

$$\frac{1}{2}\left(\alpha e^{j\beta}\right)^n u[n] \overset{Z}{\leftrightarrow} \frac{\frac{1}{2}}{1 - \alpha e^{j\beta}z^{-1}}, \quad |z| > |\alpha|$$

$$\frac{1}{2}\left(\alpha e^{-j\beta}\right)^n u[n] \overset{Z}{\leftrightarrow} \frac{\frac{1}{2}}{1 - \alpha e^{-j\beta}z^{-1}}, \quad |z| > |\alpha| \tag{7.89}$$

as

$$\begin{aligned}
F(z) &= \frac{\frac{1}{2}}{1 - \alpha e^{j\beta}z^{-1}} + \frac{\frac{1}{2}}{1 - \alpha e^{-j\beta}z^{-1}} \\
&= \frac{1 - \alpha \cos(\beta)z^{-1}}{1 - 2\alpha \cos(\beta)z^{-1} + \alpha^2 z^{-2}}, \quad |z| > |\alpha|
\end{aligned} \tag{7.90}$$

Exercise $f[n] = \sin(\beta n)u[n] \rightarrow F(z) = ?$

Example 7.17 Calculate the Z transform of

$$f[n] = \begin{cases} 1 & 0 \le n < N - 1 \\ 0 & \text{otherwise} \end{cases} \tag{7.91}$$

Solution 7.17 $f[n]$ can be written in terms of the unit step signals as

$$f[n] = u[n] - u[n - N] \tag{7.92}$$

The Z transform of (7.92) can be calculated using the transform pair

$$a^n u[n] \overset{Z}{\leftrightarrow} \frac{1}{1 - \alpha z^{-1}}, \quad |z| > |\alpha| \tag{7.93}$$

as

$$\begin{aligned} F(z) &= \frac{1}{1 - z^{-1}} - \frac{z^{-N}}{1 - z^{-1}} \\ &= \frac{1 - z^{-N}}{1 - z^{-1}}, |z| > 1 \end{aligned} \tag{7.94}$$

Example 7.18 If $h[n] = f[n] * g[n]$, then show that $H(z) = F(z)G(z)$.

Solution 7.18 The convolution expression

$$h[n] = f[n] * g[n] \tag{7.95}$$

can be written explicitly as

$$h[n] = \sum_{k=-\infty}^{\infty} f[k]g[n-k] \tag{7.96}$$

and using (7.96) in

$$H(z) = \sum_{n=-\infty}^{\infty} h[n]z^{-n} \tag{7.97}$$

we obtain

$$H(z) = \sum_{n=-\infty}^{\infty} \sum_{k=-\infty}^{\infty} f[k]g[n-k]z^{-n} \tag{7.98}$$

where letting $m = n - k$, we get $n = m + k$, and using $n = m + k$ in (7.98), we obtain

$$H(z) = \sum_{n=-\infty}^{\infty} \sum_{k=-\infty}^{\infty} f[k]g[m]z^{-(m+k)} \tag{7.99}$$

where substituting $z^{-m}z^{-k}$ for $z^{-(m+k)}$ and rearranging the equation, we get

$$H(z) = \underbrace{\sum_{k=-\infty}^{\infty} f[k]z^{-k}}_{F(z)} \underbrace{\sum_{m=-\infty}^{\infty} g[m]z^{-m}}_{G(z)} \qquad (7.100)$$

Example 7.19 Given that $y[n] = x[Mn]$ where M is a positive integer, find the Z transform of $y[n]$ in terms of the Z transform of $x[n]$.

Solution 7.19 The Z transform of $y[n]$ can be calculated as

$$Y(z) = \sum_{n=-\infty}^{\infty} y[n]z^{-n}$$

where substituting $x[Mn]$ for $y[n]$, we get

$$Y(z) = \sum_{n=-\infty}^{\infty} x[Mn]z^{-n}$$

where using $k = Mn$, we get

$$Y(z) = \sum_{k=-\infty}^{\infty} x[k]z^{-\frac{k}{M}} \qquad (7.101)$$

When (7.101) is inspected, we conclude that

$$Y(z) = X\left(z^{\frac{1}{M}}\right) \qquad (7.102)$$

Example 7.20 Given that $y[n] = x\left[\frac{n}{L}\right]$ where L is a positive integer, find the Z transform of $y[n]$ in terms of the Z transform of $x[n]$.

Solution 7.20 The Z transform of $y[n]$ can be calculated as

$$Y(z) = \sum_{n=-\infty}^{\infty} y[n]z^{-n}$$

where substituting $x\left[\frac{n}{L}\right]$ for $y[n]$, we get

$$Y(z) = \sum_{n=-\infty}^{\infty} x\left[\frac{n}{L}\right]z^{-n}$$

where using $k = \frac{n}{L}$, we get

$$Y(z) = \sum_{k=-\infty}^{\infty} x[k]z^{-kL} \qquad (7.103)$$

When (7.103) is inspected, we conclude that

$$Y(z) = X(z^L)$$

Example 7.21 Calculate the Z transform of

$$f[n] = \sum_{k=0}^{\infty} \delta[n - kM] \qquad (7.104)$$

Solution 7.21 Z transform of $f[n]$ can be calculated as

$$
\begin{aligned}
F(z) &= \sum_{n=-\infty}^{\infty} f[n]z^{-n} \\
&= \sum_{n=-\infty}^{\infty} \sum_{k=0}^{\infty} \delta[n - kM]z^{-n} \\
&= \sum_{n=-\infty}^{\infty} (\delta[n] + \delta[n - M] + \delta[n - 2M] + \cdots)z^{-n} \\
&= 1 + z^{-M} + z^{-2M} + \cdots \\
&= \frac{1}{1 - z^{-M}}
\end{aligned}
$$

Example 7.22 The relationship between the input and output of a linear and time-invariant system is described by the difference equation

$$y[n] - 0.5y[n-1] = x[n] + x[n-1] \qquad (7.105)$$

(a) Find the transfer function of the system.
(b) Find the impulse response of the system.
(c) Find the unit step response of the system.

Solution 7.22
   (a) Taking the Z transform of both sides of

$$y[n] - 0.5y[n-1] = x[n] + x[n-1] \qquad (7.106)$$

we get

$$Y(z) - 0.5z^{-1}Y(z) = X(z) + z^{-1}X(z) \tag{7.107}$$

from which we calculate the transfer function as

$$\begin{aligned} H(z) &= \frac{Y(z)}{X(z)} \\ &= \frac{1 + z^{-1}}{1 - 0.5z^{-1}} \end{aligned} \tag{7.108}$$

(b) For $x[n] = \delta[n]$, (7.106) takes the form

$$h[n] - 0.5h[n-1] = \delta[n] + \delta[n-1] \tag{7.109}$$

where taking the Z transform of both sides, we get

$$H(z) = \frac{1 + z^{-1}}{1 - 0.5z^{-1}} \tag{7.110}$$

which equals (7.108). Since practical systems have causality property, to calculate the inverse Z transform of (7.110), we can use the property

$$\alpha^n u[n] \overset{Z}{\leftrightarrow} \frac{1}{1 - \alpha z^{-1}}, \quad |z| > |\alpha| \tag{7.111}$$

Employing (7.111), we find the inverse Z transform of (7.110) as

$$h[n] = 0.5^n u[n] + 0.5^{(n-1)} u[n-1] \tag{7.112}$$

(c) The relationship between the impulse response and step response of a linear and time-invariant system is given as

$$s[n] = \sum_{k=-\infty}^{n} h[k] \tag{7.113}$$

which can be calculated using the impulse response found in the previous part as

$$\begin{aligned} s[n] &= \sum_{k=-\infty}^{n} \left(0.5^k u[k] + 0.5^{(k-1)} u[k-1] \right) \\ &= \frac{1 - 0.5^{n+1}}{1 - 0.5} + \frac{1 - 0.5^n}{1 - 0.5} \\ &= 4 - 3 \times 0.5^n, \quad n > 0 \end{aligned}$$

Thus, step response is found as $s[n] = (4 - 3 \times 0.5^n) u[n]$.

Example 7.23 The impulse response of a linear and time-invariant system is given as

$$h[n] = 5(0.2^n)u[n] + 2(0.3^n)u[n] \qquad (7.114)$$

Find the transfer function of this system.

Solution 7.23 The transfer function of the system can be obtained by evaluating the Z transform of the impulse response as

$$H(z) = \frac{5}{1 - 0.2z^{-1}} + \frac{2}{1 - 0.3z^{-1}}, \quad |z| > 0.3 \qquad (7.115)$$

Example 7.24 The relationship between the input and the output of a linear and time-invariant system is given as

$$y[n] - y[n-2] = x[n] + x[n-1] \qquad (7.116)$$

Find the transfer function of the system.

Solution 7.24 If we evaluate the Z transform of both sides of

$$y[n] - y[n-2] = x[n] + x[n-1] \qquad (7.117)$$

we obtain

$$Y(z) - z^{-2}Y(z) = X(z) + z^{-1}X(z) \qquad (7.118)$$

from which the transfer function is calculated as

$$H(z) = \frac{Y(z)}{X(z)} \rightarrow H(z) = \frac{1 + z^{-1}}{1 - z^{-2}} \qquad (7.119)$$

Example 7.25 The impulse response of a linear and time-invariant system is given as

$$h[n] = \left\{ \underbrace{0}_{n=0}, \ 1, \ 2 \right\} \qquad (7.120)$$

Find the system output for the input sequence

$$x[n] = \left\{ \underbrace{1}_{n=0}, \quad 2, \quad -1, \quad -2, \quad 1, \quad 2 \right\}$$

Solution 7.25 The output of the system can be calculated using

$$y[n] = x[n] * h[n] \tag{7.121}$$

whose Z transform can be written as

$$Y(z) = X(z)H(z) \tag{7.122}$$

The Z transforms of $x[n]$ and $h[n]$ can be calculated as

$$X(z) = \sum_{n=-\infty}^{\infty} x[n]z^{-n} \rightarrow X(z) = 1 + 2z^{-1} - z^{-2} - 2z^{-3} + z^{-4} + 2z^{-5}$$

$$H(z) = \sum_{n=-\infty}^{\infty} h[n]z^{-n} \rightarrow H(z) = z^{-1} + 2z^{-2}$$

Then, we can calculate $Y(z)$ as

$$\begin{aligned} Y(z) &= X(z)H(z) \\ &= z^{-1} + 4z^{-2} + 3z^{-3} - 4z^{-4} - 3z^{-5} + 4z^{-6} + 4z^{-7} \end{aligned} \tag{7.123}$$

If the $Y(z)$ expression in (7.123) is compared to

$$Y(z) = \sum_{n=-\infty}^{\infty} y[n]z^{-n}$$

we find that

$$y[n] = \left\{ \underbrace{0}_{n=0}, \quad 1, \quad 4. \quad 3, \quad -4, \quad -3, \quad 4, \quad 4 \right\}$$

Example 7.26 $F(z) = \frac{1}{1 - 0.8z^{-1}} \rightarrow f[n] = ?$

Solution 7.26 Using

$$\sum_{n=0}^{\infty} r^n = \frac{1}{1-r}, \quad |r| < 1 \tag{7.124}$$

we can write

$$\frac{1}{1-0.8z^{-1}} = \sum_{n=0}^{\infty} (0.8z^{-1})^n$$

leading to

$$\frac{1}{1-0.8z^{-1}} = \sum_{n=0}^{\infty} 0.8^n z^{-n} \qquad (7.125)$$

When (7.125) is compared to

$$F(z) = \sum_{n=-\infty}^{\infty} f[n]z^{-n} \qquad (7.126)$$

we find that

$$f[n] = 0.8^n u[n]$$

Example 7.27 $F(z) = \frac{1}{z-\frac{1}{4}} \to f[n] = ?$ $|z| > \frac{1}{4}$.

Solution 7.27 $F(z)$ can be written as

$$F(z) = \frac{z^{-1}}{1-\frac{1}{4}z^{-1}} \to F(z) = z^{-1}\left(\frac{1}{1-\frac{1}{4}z^{-1}}\right)$$

where the inverse Z transform of the term inside the parenthesis is $\left(\frac{1}{4}\right)^n u[n]$. Multiplying by z^{-1} in Z domain corresponds to shifting in the time domain, then we get

$$f[n] = \left(\frac{1}{4}\right)^{n-1} u[n-1]$$

Exercise $F(z) = \frac{1}{z+1.5} \to f[n] = ?$

Exercise The transfer function of a linear and time-invariant system is given as

$$H(z) = \frac{2z^2 - 3z}{z^2 + 0.5z - 1}$$

Find the relationship between the input and the output of the system.

Example 7.28 $g[n] = \alpha^{-n}u[-n] \rightarrow G(z) = ?$

Solution 7.28 We use the property

$$f^*[-n] \overset{Z}{\leftrightarrow} F^*\left(\frac{1}{z^*}\right), \quad \text{ROC}_1 = \frac{1}{\text{ROC}} \tag{7.127}$$

for the solution of the example. Let $f[n] = \alpha^n u[n]$ and $g[n] = f[-n]$, then we have

$$G(z) = F(z^{-1}) \tag{7.128}$$

The Z transform of $f[n] = \alpha^n u[n]$ is

$$F(z) = \frac{1}{1 - \alpha z^{-1}}, \quad |z| > |\alpha| \tag{7.129}$$

and since $g[n] = f[-n]$, the Z transform of $g[n]$ can be written

$$G(z) = \frac{1}{1 - \alpha z}, \quad \left|\frac{1}{z}\right| > |\alpha| \tag{7.130}$$

which can be rearranged as

$$G(z) = \frac{-\alpha^{-1}z^{-1}}{1 - \alpha^{-1}z^{-1}}, \quad |z| < |\alpha|^{-1} \tag{7.131}$$

Example 7.29 The relationship between $g[n]$ and $f[n]$ is given as

$$g[n] = \sum_{k=-\infty}^{n} f[k] \tag{7.132}$$

Express the Z transform of $g[n]$ in terms of the Z transform of $f[n]$.

Solution 7.29 The expression

$$g[n] = \sum_{k=-\infty}^{n} f[k]$$

can be written as

$$g[n] = \sum_{k=-\infty}^{\infty} f[k]u[n-k] \tag{7.133}$$

where the upper limit is ∞ different than in (7.132). It is seen from (7.133) that

$$g[n] = f[n] * u[n] \tag{7.134}$$

whose Z transform can be written as

$$G(z) = F(z)U(z) \tag{7.135}$$

leading to

$$G(z) = \frac{F(z)}{1 - z^{-1}} \tag{7.136}$$

Example 7.30 $F(z) = \frac{z^2}{(z+1)(z-2)}$, $|z| > 2 \rightarrow f[n] = ?$

Solution 7.30 The numerator and denominator of the rational polynomial have equal degrees, and the first thing that comes to mind is to divide the polynomial. However, we can solve the problem in a simpler way as follows. We first write

$$\frac{F(z)}{z}$$

as

$$\frac{F(z)}{z} = \frac{z}{(z+1)(z-2)}$$

which can be expanded as

$$\frac{F(z)}{z} = \frac{1}{3}\frac{1}{z+1} + \frac{2}{3}\frac{1}{z-2}$$

leading to

$$F(z) = \frac{1}{3}\frac{z}{z+1} + \frac{2}{3}\frac{z}{z-2}$$

which can be written as

$$F(z) = \frac{1}{3}\frac{1}{1+z^{-1}} + \frac{2}{3}\frac{1}{1-2z^{-1}}$$

whose inverse Z transform can be obtained as

$$f[n] = \left(\frac{(-1)^n + 2^{n+1}}{3}\right)u[n]$$

7.4   Solving the Discrete-Time Difference Equations of Causal Systems Using Z Transform

A causal $f[n]$ signal satisfies the condition $f[n] = 0$, $n < 0$. For a non-causal signal $f[n]$, we have the Z transform pair

$$f[n + n_0] \overset{Z}{\leftrightarrow} z^{n_0} F(z) \tag{7.137}$$

First of all, let us give the Z transformation properties that are used for the solution of the causal discrete-time difference equations, and then let us give an example to explain the subject.

Properties We have the transform pairs

$$f[n + 1] \overset{Z}{\leftrightarrow} zF(z) - zf[0] \tag{7.138}$$

$$f[n + 2] \overset{Z}{\leftrightarrow} z^2 F(z) - z^2 f[0] - zf[1] \tag{7.139}$$

and in general, we have

$$f[n + m] \overset{Z}{\leftrightarrow} z^m\left(F(z) - \sum_{k=0}^{m-1} f[k]z^{m-k}\right) \tag{7.140}$$

which can be solved for the solution of the difference equations.

Example 7.31 Find the solution of the Fibonacci equation

$$f[n] + f[n + 1] = f[n + 2], \quad n \le 1 \tag{7.141}$$

using the initial conditions $f[0] = f[1] = 1$.

Solution 7.31 The Z transform of both sides of

$$f[n] + f[n+1] = f[n+2]$$

can be evaluated using the transform pairs

$$f[n+1] \overset{Z}{\leftrightarrow} zF(z) - zf[0] \quad f[n+2] \overset{Z}{\leftrightarrow} z^2F(z) - z^2f[0] - zf[1]$$

as

$$F(z) + zF(z) - z = z^2F(z) - z^2 - z$$

from which we get $F(z)$ as

$$F(z) = \frac{z^2}{z^2 - z - 1}$$

which can be written as

$$F(z) = \left(\frac{\alpha}{\alpha + \frac{1}{\alpha}}\right)\frac{z}{z - \alpha} + \left(\frac{\frac{1}{\alpha}}{\alpha + \frac{1}{\alpha}}\right)\frac{z}{z + \frac{1}{\alpha}} \tag{7.142}$$

where $\alpha = \frac{1+\sqrt{5}}{2}$ called golden ratio. The inverse Z transform of (7.142) can be calculated as

$$f[n] = \left(\frac{\alpha}{\alpha + \frac{1}{\alpha}}\right)\alpha^n u[n] + \left(\frac{\frac{1}{\alpha}}{\alpha + \frac{1}{\alpha}}\right)\left(-\frac{1}{\alpha}\right)u[n] \tag{7.143}$$

Summary

The Z transform of $f[n]$ is calculated using

$$F(z) = \sum_{n=-\infty}^{\infty} f[n]z^{-n} \tag{7.144}$$

where $F(z)$ converges for a set of z values. The region of these z values on the complex plane is called the region of convergence of $F(z)$. Inverse Z transform is calculated using

$$f[n] = \frac{1}{2\pi j} \oint F(z) z^{n-1} dz \qquad (7.145)$$

which is not used too much in practice; instead, the table of known Z transform pairs is used for the calculation of inverse Z transform.

Let $F(z)$ and $G(z)$ be the Z transforms of $f[n]$ and $g[n]$ with convergence regions ROC_f and ROC_g. The Z transform properties are given in Table 7.1, and all these properties can be proved using the definition of Z transform.

The Z transforms of some known basic signals are summarized in Table 7.2. Using this table, Z transforms of signals or inverse Z transforms of polynomials can be calculated.

Table 7.1 Z Transform properties

Signal	Z transform	Convergence region
$af[n] + bg[n]$	$aF(z) + bG(z)$	$\text{ROC} = \text{ROC}_f \cap \text{ROC}_g$
$f[-n]$	$F\left(\frac{1}{z}\right)$	$\text{ROC} = \frac{1}{\text{ROC}_f}$
$f[n - n_0]$	$z^{-n_0} F(z)$	$\text{ROC} = (\text{ROC}_f \text{ possibly excluding the points } z = 0 \text{ or } z = \infty)$
$\alpha^n f[n]$	$F\left(\frac{z}{\alpha}\right)$	$|\alpha|\text{ROC}_f$
$f[n] * g[n]$	$F(z)G(z)$	$\text{ROC} \supset (\text{ROC}_f \cap \text{ROC}_g)$
$nf[n]$	$-z \frac{dF(z)}{dz}$	$\text{ROC} = (\text{ROC}_f \text{ possibly excluding the points } z = 0 \text{ or } z = \infty)$
$e^{j\Omega_0 n} f[n]$	$F(e^{-j\Omega_0} z)$	$\text{ROC} = \text{ROC}_f$
$Re\{f[n]\}$	$\frac{1}{2}\left(F(z) + F^*(z^*)\right)$	$\text{ROC} = \text{ROC}_f$
$Im\{f[n]\}$	$\frac{1}{2j}\left(F(z) - F^*(z^*)\right)$	$\text{ROC} = \text{ROC}_f$

Table 7.2 Z transform table

Signal	Z transform	Region of convergence
$\delta[n]$	1	Entire z plane
$u[n]$	$\frac{1}{1 - z^{-1}}$	$|z| > 1$
$-u[-n - 1]$	$\frac{1}{1 - z^{-1}}$	$|z| < 1$
$\alpha^n u[n]$	$\frac{1}{1 - \alpha z^{-1}}$	$|z| > |\alpha|$
$-\alpha^n u[-n - 1]$	$\frac{1}{1 - \alpha z^{-1}}$	$|z| < |\alpha|$
$n\alpha^n u[n]$	$\frac{\alpha z^{-1}}{(1 - \alpha z^{-1})^2}$	$|z| > |\alpha|$
$-n\alpha^n u[-n - 1]$	$\frac{\alpha z^{-1}}{(1 - \alpha z^{-1})^2}$	$|z| < |\alpha|$
$[\cos(\Omega_0 n)]u[n]$	$\frac{1 - [\cos(\Omega_0)]z^{-1}}{1 - [2\cos\Omega_0]z^{-1} + z^{-2}}$	$|z| > 1$
$[\sin(\Omega_0 n)]u[n]$	$\frac{1 - [\sin(\Omega_0)]z^{-1}}{1 - [2\cos\Omega_0]z^{-1} + z^{-2}}$	$|z| > 1$
$[\alpha^n \cos(\Omega_0 n)]u[n]$	$\frac{1 - [\alpha\cos(\Omega_0)]z^{-1}}{1 - [2\alpha\cos\Omega_0]z^{-1} + \alpha^2 z^{-2}}$	$|z| > |\alpha|$
$[\alpha^n \sin(\Omega_0 n)]u[n]$	$\frac{1 - [\alpha\sin(\Omega_0)]z^{-1}}{1 - [2\alpha\cos\Omega_0]z^{-1} + \alpha^2 z^{-2}}$	$|z| > |\alpha|$

Problems

1. Find the Z transforms, and determine the convergence regions of the signals:

   (a) $f[n] = (-1)^n u[n-3] + u[n-1]$
   (b) $f[n] = (-2)^n u[-n-2]u[-n-1]$
   (c) $f[n] = -2\delta[n-2] + 3\delta[n+1]$
   (d) $f[n] = (-0.4)^n(n-1)u[n-2]$

2. Find the inverse Z transform of the signals:

   (a) $F(z) = \cos(z), \ |z| > 1$
   (b) $F(z) = \frac{z-2}{1-z^{-1}}, \ |z| > 1$
   (c) $F(z) = \frac{z}{1-0.2z^{-2}}, \ |z| > 0.2$
   (d) $F(z) = \frac{1}{(1-0.2z^{-1})(1-0.4z^{-1})}, \ |z| > 0.4$
   (e) $F(z) = z^2 - \frac{1}{2}z - 1 + \frac{1}{2}z^{-1}$
   (f) $F(z) = \log(1 + z^{-1}), \ |z| > 1$

3. The relationship between $y[n]$ and $x_1[n]$, $x_2[n]$ is given as

$$y[n] = x_1[n+3] * x_2[-n+1]$$

   where $x_1[n] = \left(\frac{1}{2}\right)^n u[n]$ and $x_2[n] = \left(\frac{1}{3}\right)^n u[n]$. We have the transform pair

$$a^n u[n] \overset{Z}{\leftrightarrow} \frac{1}{1-az^{-1}}, \ |z| > |a|$$

   Find $Y(z)$.

4. The relationship between the input and output of a linear and time-invariant system is described by the difference equation

$$y[n-1] - \frac{5}{3}y[n] + y[n+1] = x[n]$$

   The system has the stability property. Find the unit step response of this system.

5. $f[n] = \left(\frac{1}{3}\right)^n \sin\left(\frac{\pi}{4}n\right)u[n] \rightarrow F(z) = ?$

6. Find the solution of the difference equation with the given initial condition

$$f[n+1] - 2f[n] = (-1)^n, \ f[0] = -2$$

   using Z transform.

7. $F(z) = \frac{z}{z^2+1} \rightarrow f[n] = ?$

8. The digital signal $x[n]$ is defined as

$$x[n] = \begin{cases} 1 & 0 \le n < 6 \\ 0 & \text{otherwise} \end{cases}$$

    (a) Write $x[n]$ in terms of the unit step signals.
    (b) Find the Z transform of $x[n]$.
    (c) If $x[n] = \sum_{k=-\infty}^{n} g[k]$, write $g[n]$ in terms of $x[n]$.
    (d) Find the Z transform of $g[n]$.

9. The transfer functions of two linear time-invariant and stable systems are given
below. Without using the inverse Z transform, determine whether these systems
are causal or not:

    (a) $H_1(z) = \dfrac{z^{-1} - \frac{1}{2}z^{-2}}{1 + \frac{1}{2}z^{-1} - \frac{3}{16}z^{-2}}$

    (b) $H_2(z) = \dfrac{1 - \frac{4}{3}z^{-1} + \frac{1}{3}z^{-2}}{z^{-1}\left(1 - \frac{1}{2}z^{-1}\right)\left(1 - \frac{1}{3}z^{-1}\right)}$

10. Let us denote the even and odd parts of $x[n]$ by $x_e[n]$ and $x_o[n]$, respectively, and
we have $x[n] = x_e[n] + x_o[n]$. Express the Z transforms of $x_e[n]$ and $x_o[n]$ in
terms of $X(z)$.

11. Find the Z transforms of the following signals, determine the regions of convergences, and decide whether the Fourier transforms can be calculated for the
given signals:

    (a) $x[n] = \left(\frac{1}{4}\right)^n (u[n+5] - u[n-3])$
    (b) $x[n] = |n| \left(\frac{1}{4}\right)^{|n|}$
    (c) $x[n] = 2^n \sin\left(\frac{\pi}{4}n + \frac{\pi}{6}\right) u[-n-2]$

12. Find the inverse Z transform of

$$H(z) = \frac{1 - z^{-2}}{(1 - z^{-1})(1 + 0.5z^{-1})^2 (1 - 0.8z^{-1})^3}$$

Chapter 8
Practical Applications

In this chapter, we consider a practical topic as an application of the topics we covered in the previous chapters. Although students think of many of the signal processing subjects they learn in theoretical courses only as mathematical expressions, almost all of them have application areas. In this chapter, we first consider an example application of the subjects we learned. Secondly, we will describe the development of the fast Fourier transform algorithm as an example for scientific work.

Formulas or algorithms that require too much processing are impractical for real-time applications since they consume too much processing time for computer implementations. Today's computers can run faster than their old counterparts. Algorithms, which were considered useless 30 or 40 years ago for computer implementations due to large latency, are today used in practical applications. Scientists are constantly trying to achieve improved techniques by reducing the complexity of invented techniques. Today, many countries are in a position to produce mobile phones. But applying rough technology alone is not enough for the device produced to hold on to the market. Devices produced with improved techniques may bring one to an advantageous position compared to the original owners of the technique.

After the Second World War, with the use of the first computer, many electronic circuits, which were mainly designed as hardware, began to be designed as software. Although hardware-designed circuits run faster than software ones, recent advances in microprocessor technology have caused software-intensive designs to come to the forefront compared to hardware-intensive ones. With the prominence of software, researchers have worked to reduce the amount of mathematical computation required by existing algorithms. Algorithms whose complexities are reduced by scientific methods are used in practical applications. Discrete Fourier transform is one of the most important topics of signal processing, and it is used in many practical applications. Fast Fourier transform Goertzel and Chirp algorithms, which require less processing, have been developed to calculate the discrete Fourier transform faster. In this chapter, we explain Goertzel and fast Fourier transform algorithms. We consider the detection of dual-tone multi-frequencies using the Goertzel algorithm.

O. Gazi, *Principles of Signals and Systems*, https://doi.org/10.1007/978-3-031-17789-7_8

8.1   Computational Complexity of the Discrete Fourier Transform Algorithm and Complexity Reduction

Computational Complexity

For the computation of

$$x^2 + xy \tag{8.1}$$

we need to perform:

1. $x^2 = xx$, one multiplication
2. xy, one multiplication
3. $x^2 + xy$, one addition

Thus, in total, two multiplications and one addition, i.e., three operations are performed.

The expression $x^2 + xy$ can be written as

$$x(x + y) \tag{8.2}$$

and for the computation of (8.2), we need:

1. $x + y$, one addition
2. $x(x + y)$, one multiplication

Hence, two operations are needed for the calculation of (8.2).

It is seen that we can calculate the same mathematical expression with fewer operations. A similar logic is used for the calculation of discrete Fourier transform. Discrete Fourier transform is calculated using

$$F[k] = \frac{1}{\sqrt{N}} \sum_{N=0}^{N-1} f[n] e^{-j\frac{2\pi}{N}kn}, \quad k = 0,1,\ldots,N-1 \tag{8.3}$$

where $f[k]$ is a complex digital sequence. For the easiness of the notation, let us use $e_n = e^{-j\frac{2\pi}{N}}$ in (8.3), and then we get

$$F[k] = \frac{1}{\sqrt{N}} \sum_{N=0}^{N-1} f[n] e^{-j\frac{2\pi}{N}kn}, \quad k = 0,1,\ldots,N-1 \tag{8.4}$$

where for each k value there are N complex multiplication operations and $N-1$ complex addition operations. For all the k values, $0 \leq k < N$, there are N^2 complex multiplication and $N(N-1)$ complex addition operations. The total number of arithmetic operations is $N^2 + N(N-1)$ which can be approximated as $2N^2$ for very large N values. Thus we can conclude that the total amount of computation for the calculation of the Fourier transform is proportional to N^2.

The amount of calculation required for the calculation of the Fourier transform can be reduced. The complex signals $f[n]$ and e_N^{kn} appearing in (8.4) can be written as

$$f[n] = \text{Re}\left\{f[n]\right\} + j\text{Im}\{f[n]\} \quad e_N^{kn} = \text{Re}\left\{e_N^{kn}\right\} + j\text{Im}\{e_N^{kn}\} \tag{8.5}$$

where $\text{Re}\{\cdot\}$ and $\text{Im}\{\cdot\}$ indicate the real and imaginary parts of the complex expressions. Substituting (8.5) into (8.4), we obtain

$$
\begin{aligned}
F[k] = \frac{1}{\sqrt{N}} \sum_{N=0}^{N-1} & \text{Re}\left\{f[n]\right\} \text{Re}\left\{e_N^{kn}\right\} - \text{Im}\{f[n]\}\text{Im}\{e_N^{kn}\} \\
& + j\left(\text{Re}\left\{f[n]\right\}\text{Im}\{e_N^{kn}\} + \text{Im}\{f[n]\}\text{Re}\left\{e_N^{kn}\right\}\right), \quad k = 0,1,\ldots,N-1
\end{aligned}
\tag{8.6}
$$

In (8.6), for each value of k, the number of multiplications is $4N$, and the number of additions is $4N - 2$.

The number of addition operations required for $n = 0$ is 2. For other n values, namely, $n = 1, 2, \ldots$, the number of additions is 2 for the complex number of the current n value, but this complex number is added to the previous summations; hence, the total number of addition operations becomes 4. If we consider all k values, there are $4N$ addition operations.

However, since there is no complex number before $n = 0$, the complex number at $n = 0$ is not added with the previous one; hence, the total number of addition operations is $4N - 2$.

To reduce the amount of total computation, the properties of the exponential signal e_N^{kn} can be used. The exponential signal e_N^{kn} has the properties

$$e_N^{kn} = e_N^{k(n+N)} \quad e_N^{-kn} = e_N^{k(N-n)} \quad e_N^{kN} = 1 \tag{8.7}$$

where e_N^{kn} equals to 1 for $n = N$, since we have

$$
\begin{aligned}
e^{-j\frac{2\pi}{N}kN} &= e^{-j2\pi k} \\
&= \underbrace{\cos\left(2\pi k\right)}_{=1} - \underbrace{j\sin\left(2\pi k\right)}_{=0} \\
&= 1 - 0 \\
&= 1.
\end{aligned}
$$

The summation expression in

$$F[k] = \frac{1}{\sqrt{N}} \sum_{N=0}^{N-1} \mathrm{Re}\,\{f[n]\}\,\mathrm{Re}\,\{e_N^{kn}\} - \mathrm{Im}\{f[n]\}\mathrm{Im}\{e_N^{kn}\}$$

$$+j\left(\mathrm{Re}\,\{f[n]\}\mathrm{Im}\{e_N^{kn}\} + \mathrm{Im}\{f[n]\}\,\mathrm{Re}\,\{e_N^{kn}\} + \mathrm{Im}\{f[n]\}\,\mathrm{Re}\,\{e_N^{kn}\}\right)$$

(8.8)

contains terms for both n and $N-n$; for instance, for $n=2$ a term for $n=N-2$ also appears in (8.8), and the terms for n and $N-n$ can be combined as

$$\mathrm{Re}\,\{f[n]\}\,\mathrm{Re}\,\{e_N^{kn}\} + \mathrm{Re}\,\{f[N-n]\}\,\mathrm{Re}\,\left\{e_N^{k(N-n)}\right\}$$

$$= [\,\mathrm{Re}\,\{f[n]\} + \mathrm{Re}\,\{f[N-n]\}]\,\mathrm{Re}\,\{e_N^{kn}\}$$

$$\mathrm{Im}\{f[n]\}\mathrm{Im}\{e_N^{kn}\} + \mathrm{Im}\{f[N-n]\}\mathrm{Im}\left\{e_N^{k(N-n)}\right\}$$

$$= [\mathrm{Im}\{f[n]\} + \mathrm{Im}\{f[N-n]\}]\,\mathrm{Re}\,\{e_N^{kn}\}$$

(8.9)

where the number of multiplication operations is halved.

Since the expression e_N^{kn} is a sinusoidal expression, it equals 0 or 1 for some n values. These values do not need to be multiplied, which contributes to reducing the amount of processing

Fast Fourier transform was developed to calculate the discrete Fourier transform quickly. When the frequency range is taken from 0 to 2π, the Goertzel algorithm is more efficient than the fast Fourier transform.

8.2 GoertzelAlgorithm

The Goertzel algorithm was invented in 1958 to decrease the computational complexity of the discrete Fourier transform algorithm. This algorithm is used to detect dual-tone multi-frequencies, DTMF, used in telephones.

Now let us explain the derivation of the algorithm. Discrete Fourier transform is calculated using

$$F[k] = \frac{1}{\sqrt{N}} \sum_{n=0}^{N-1} f[n] e^{-j\frac{2\pi}{N}kn}$$

(8.10)

We have $e_N^{-kN} = 1$ Multiplying both sides of (8.10) by e_N^{-kN}, we get

$$F[k] = \frac{e_N^{-kN}}{\sqrt{N}} \sum_{n=0}^{N-1} f[n] e_N^{kn}$$

(8.11)

which can be written as

$$F[k] = \frac{1}{\sqrt{N}} \sum_{n=0}^{N-1} f[n] e_N^{-k(N-m)} \tag{8.12}$$

where changing the parameter n to m, we obtain

$$F[k] = \frac{1}{\sqrt{N}} \sum_{m=0}^{N-1} f[m] e_N^{-k(N-m)} \tag{8.13}$$

where using $f[n] = 0$, $n < 0$, and $n > N$, we can write (8.13) as

$$F[k] = \frac{1}{\sqrt{N}} \left(f[n] * \left[e_N^{-kn} u[n] \right] \right) \big|_{n=N} \tag{8.14}$$

The signals $h[n]$ and $y_k[n]$ are defined as

$$\begin{aligned} h[n] &= e_N^{-kn} u[n] \\ y_k[n] &= f[n] * h[n] \end{aligned} \tag{8.15}$$

Equation (8.13) can be written in terms of $y_k[n]$ as

$$F[k] = \frac{1}{\sqrt{N}} (y_k[n]) \bigg|_{n=N} \tag{8.16}$$

The output of the linear and time-invariant system with impulse response $h[n] = e_N^{-kn} u[n]$ is $y_k[n]$ which is shown in Fig. 8.1.

The Z transform of the impulse response of Fig. 8.1 can be calculated using $H(z) = \sum_{n=-\infty}^{\infty} h[n] z^{-n}$ where substituting $h[n] = e_N^{-kn} u[n]$, we obtain

$$H(z) = \sum_{n=0}^{\infty} e_N^{-kn} z^{-n}$$

which can be rearranged as

$$H(z) = \sum_{n=0}^{\infty} \left(e_N^{-k} z^{-1} \right)^n$$

resulting in

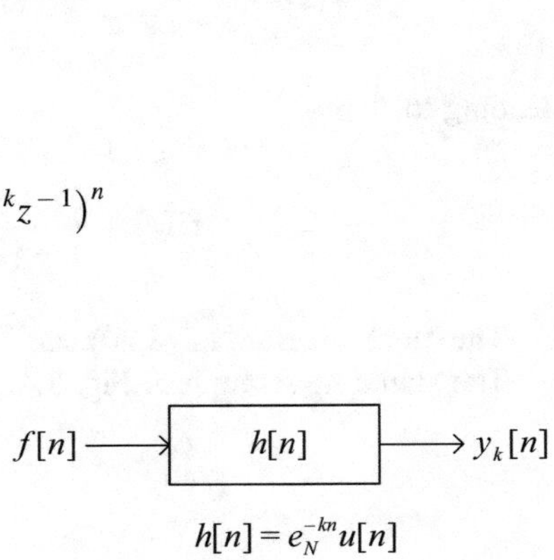

Fig. 8.1 An LTI system with impulse response $h[n] = e_N^{-kn} u[n]$

$$f[n] \longrightarrow \boxed{\quad h[n] \quad} \longrightarrow y_k[n]$$

$$h[n] = e_N^{-kn} u[n]$$

Fig. 8.2 Block diagram
representation of (8.17)

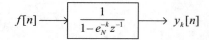

Fig. 8.3 Signal flow
diagram of Fig. 8.2

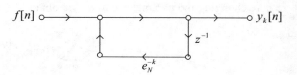

Fig. 8.4 The block diagram
of (8.20)

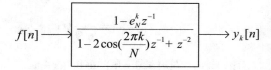

$$H(z) = \frac{1}{1 - e_N^{-k} z^{-1}} \tag{8.17}$$

Using (8.17), we can draw the system block as in Fig. 8.2.
Signal flow diagram of Fig. 8.2 can be drawn as in Fig. 8.3.
Using Fig. 8.3, we write the equation

$$y_k[n] = e_N^{-k} y_k[n-1] + f[n] \tag{8.18}$$

where $y_k[-1] = 0$.

The recursive expression in (8.18) involves a complex multiplication expression which increases the overhead of the processes in practical applications. To alleviate the overhead created by the complex multiplication, we can manipulate the transfer function to get a more useful expression as in

$$H(z) = \frac{1}{1 - e_N^{-k} z^{-1}} \rightarrow H(z) = \frac{1 - e_N^k z^{-1}}{1 - e_N^k z^{-1}} \times \frac{1}{1 - e_N^{-k} z^{-1}} \tag{8.19}$$

leading to

$$H(z) = \frac{1 - e_N^k z^{-1}}{1 - \left(2 \cos\left(\frac{2\pi}{N} k\right)\right) z^{-1} + z^{-2}} \tag{8.20}$$

The block diagram of (8.20) can be drawn as in Fig. 8.4.
The signal flow graph of Fig. 8.4 is depicted in Fig. 8.5.

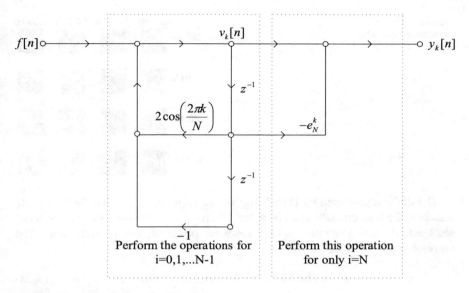

Fig. 8.5 The signal flow graph of Fig. 8.4

The direct realization of the signal flow graph in Fig. 8.5 can be written as in

$$v_k[i] = 2\cos\left(\frac{2\pi k}{N}\right) \; v_k[i-1] - v_k[i-2] + f[i], \; i = 0, 1, \ldots, N-1$$
$$y_k[i] = v_k[i] - e_N^k v_k[i-1], \; i = N \tag{8.21}$$

where initial conditions are $v_k[-2] = v_k[-1] = 0$.

In (8.21), there is one real multiplication and a single addition operation, and complex multiplication is performed for only $i = N$. Thus, equations in (8.21) are more suitable for practical applications. After calculation of $y_k[n]$ in (8.21), the Fourier series coefficients are evaluated using

$$F[k] = \frac{1}{\sqrt{N}} \left(y_k[n]\right)\big|_{n=N} \tag{8.22}$$

8.3 Dual-Tone Multi-frequency (DTMF) Detection Using Goertzel Algorithm

Dual-tone multi-frequency, DTMF, was first developed in American Bell Laboratories. Today, this system is used by many communication devices.

Fig. 8.6 The digits
represented by DTMF
signals

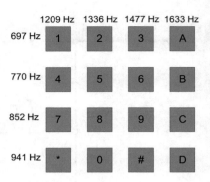

The digits represented by DTMF signals are depicted in Fig. 8.6. In the DTMF standard, the tone duration must be at least 40 ms which is taken into account when electronic devices generating DTMF tones are produced. For instance, the signal representing the digit "5" is

$$c_1 \sin (2\pi \times 770 \times t) + c_2 \sin (2\pi \times 1336 \times t) \qquad (8.23)$$

which must have a duration of at least 40 ms. The coefficients in (8.23) can be chosen as $c_1 = c_2 = \frac{1}{2}$.

On the receiver side, the key pressed on the transmitter is determined.Different methods can be used to detect the key pressed. Before the detection operation, the received tone is sampled.

The sampling frequency used for telephone speech is $F_s = 8\,\text{kHz}$. As it is mentioned before, DTMF tone duration must be at least 40 ms. The number of samples taken from a tone of 40 ms duration can be calculated as $40 \times 10^{-3} \times 8000 = 320$. The generation of the tone corresponding to the digit "1" using MATLAB is performed in Prog 8.1 where the sampling frequency is taken as 8 kHz and tone duration is chosen as 100 ms.

Prog 8.1
```
low_frequency = 697; % Low Frequency

high_frequency= 1209 % High Frequency

Sampling_Frequency = 8000; % Sampling Frequency: 8 kHz

N = 800;   % Tone Duration: 100 ms
t = (0:N-1)/ Sampling_Frequency;
two_pi_t = 2*pi*t;

% Tone Generation
tone =sin(low_frequency* two_pi_t)+sin(high_frequency* two_pi_t)

plot(t*1000,tone); % Tone Plot
axis([0 25 -2 2]);
```

```
xlabel('Time (ms)');
ylabel('Amplitude');
title(['"', num2str(1),'"': (',num2str(697),'Hz',',',num2str
(1209),'Hz',')']);
```

The tone graph corresponding to digit "1" is displayed in Fig. 8.7.

The duration of the signal generated on the transmitter side is 100 ms, which corresponds to 800 samples when the sampling frequency is chosen as 8 kHz. Not all of these 800 samples need to be used on the receiver side for tone detection. The researchers showed that the use of 205 samples at the receiver side for a sampling frequency of 8 kHz minimizes the minimum mean square detection error. In other words, it will be sufficient to use only 205 out of 800 samples on the receiver side. This means that the signal with a duration of approximately 25 ms is sufficient for the detection process.

Although the transmitter knows which signal it is sending, this information is not available to the receiver. On the receiver side, different methods can be followed to determine the key pressed. One of these methods is as follows. The absolute value of the discrete Fourier transform coefficients at the receiver is calculated for each frequency value, i.e., for each "k" value, we calculate

$$|X[k]| = \frac{1}{\sqrt{N}} \left| \sum_{N=0}^{N-1} x[n] e^{-j\frac{2\pi}{N}kn} \right|, \ 0 \le k < N \tag{8.24}$$

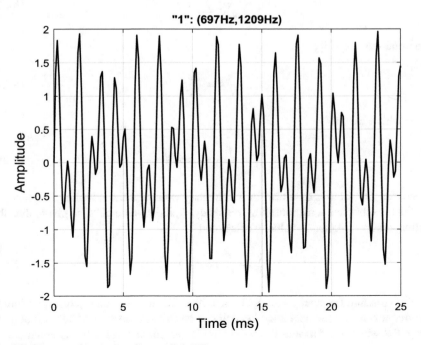

Fig. 8.7 Tone graph corresponding to digit "1"

and the numbers "k"giving the highest values and their corresponding frequencies are determined. By inspecting the selected frequencies, the key pressed is determined. However, the complexity of this method is high. Instead, the Goertzel algorithm, which has less complexity and allows us to get the results faster, is used.

The steps of tone detection with the Goertzel algorithm are as follows:

First, the "k" values corresponding to eight frequencies used on the telephone keypad are determined as follows.

If the continuous time sinusoidal signal $\sin(2\pi f t)$ is sampled with a sampling frequency of F_s, we get

$$\sin\left(n\frac{2\pi}{F_s}f\right)$$

where

$$\Omega = \frac{2\pi}{F_s}f \tag{8.25}$$

is the digital angular frequency. When (8.25) is equated to the exponential term $\Omega = \frac{2\pi}{F_s}k$ in

$$f[n] = \frac{1}{\sqrt{N}}\sum_{k=0}^{N-1}F[k]e^{j\frac{2\pi}{N}kn} \tag{8.26}$$

we obtain the equality

$$\frac{2\pi}{N}k = \frac{2\pi}{F_s}f \tag{8.27}$$

from which we get

$$\frac{k}{N} = \frac{f}{F_s} \rightarrow k = K\frac{f}{F_s} \tag{8.28}$$

where the "k" value is obtained for a frequency value and the sampling frequency is F_s. On the other hand, for sampling frequency F_s, if the value of "k" is given, then the corresponding frequency value is calculated as

$$\frac{k}{N} = \frac{f}{F_s} \rightarrow f = F_s\frac{k}{N} \tag{8.29}$$

For sampling frequency $F_s = 8$ kHz and $N = 205$, the "k" values corresponding to the frequencies on the telephone keypad can be calculated using MATLAB code in Prog. 8.2 where the formula $k = N\frac{f}{F_s}$ is implemented, and Table 8.1 is obtained.

Table 8.1 Frequency and coefficient table

Original frequency	k
697	18
770	20
852	22
941	24
1209	31
1336	34
1477	38

Table 8.2 Coefficient and estimated frequency table

k	Estimated frequency
18	702
20	780
22	859
24	937
31	1210
34	1327
38	1483

Prog. 8.2

```
low_frequencies = [697 770 852 941]; % Low Frequencies
high_frequencies = [1209 1336 1477]; % High Frequencies

Sampling_Frequency = 8000; % Sampling Frequency Fs=8 kHz
Nt = 205;
original_frequency = [low_frequencies (:);high_frequencies(:)]
% Original Frequencies

k = round((original_frequency *Nt)/ Sampling_Frequency) % DFT 'k' index
value
```

On the other hand, for the "k" values, the corresponding frequencies can be calculated using the formula

$$f = F_s \frac{k}{N} \tag{8.30}$$

and a MATLAB code is written in Prog. 8.3 for this calculation. The calculation results are depicted in Table 8.2.

Prog. 8.3

```
Sampling_Frequency = 8000;
Nt = 205;
k = [ 18 20 22 24 31 34 38]

estimated_frequency = round(k* Sampling_Frequency /Nt)
```

1. In the second step, for each "k" value found in the first step, the discrete Fourier transform coefficients at the receiver are calculated using the Goertzel algorithm.
2. In the third step, the absolute value of the discrete Fourier transform calculated for each "k" value is checked, and the two largest "k" values are recorded. Then, the frequencies corresponding to the recorded "k" values are found, and the numbers corresponding to these two frequencies on the keypad are determined. Let's explain how the tone corresponding to the "1" is detected using the third step. In order to find out which key was pressed, the sampled data atthe receiver is passed through the Goertzel algorithm, the frequencies of the sinusoidal signals in the data are found, and finally the key pressed is determined. In MATLAB Prog. 8.4, using 205 samples, the transmitted frequencies are determined.

Prog. 8.4

```
low_frequency = 697;   % Low Frequency
high_frequency= 1209;  % High Frequency

Sampling_Frequency = 8000; % Sampling Frequency 8 kHz
N = 800;    % 100ms Duration Tone
t = (0:N-1)/ Sampling_Frequency; % two_pi_t = 2*pi*t;

% Tone Generation
tone =sin(low_frequency*two_pi_t)+sin(high_frequency*two_pi_t);
k=[18 20 22 24 31 34 38]; % Frequency coefficients to check in detection
Nt = 205; % For 8kHz Sampling frequency, It is sufficient to use
205 samples
tone= tone(1:Nt);

ydft = goertzel(tone,k+1); % In Goertzel's Algorithm, Index starts from
1, not 0

% Frequencies corresponding to k values
estimated_frequency = round(k*Sampling_Frequency/Nt);

% DFT Absolute value for different frequencies
stem(estimated_frequency,abs(ydft));

title(['"', num2str(1),'": (',num2str(697),'Hz',',',num2str
(1209),'Hz',')']);
set(gca, 'XTick', estimated_frequency, 'XTickLabel',
estimated_frequency, 'Xlim', [650 1550]);
ylabel('DFT Amplitude');
xlabel('Frequency (Hz)');
```

The graph obtained using Prog. 8.4 is depicted in Fig. 8.8. If we examine this graph carefully, we see that the absolute value of the discrete Fourier transform takes large values for $k = 18$ and $k = 31$.

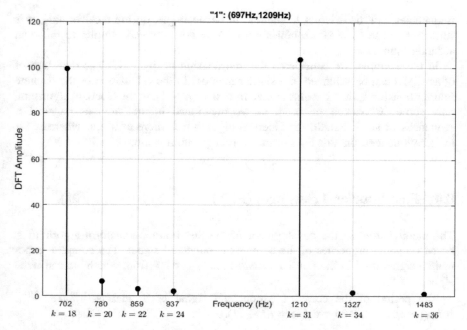

Fig. 8.8 Plot of absolute DFT values for different frequencies

The frequencies corresponding to these "k" values are 702 and 1210, and the frequencies closest to these values on the keypad are 697 and 1209, which corresponds to the digit "1" on the keypad.

A question may come to the reader's mind as follows:

Question:
"We already know the frequencies corresponding to the 'k' values for the keypad, so why are we calculating approximate frequency values again."

Answer:
We explained the general situation. If the communication system had used more frequencies and not just the frequencies on the keypad, then it would be impractical to store all frequencies in memory corresponding to the "k" value for high N values.

For these reasons, the frequencies corresponding to the "k" values determined as a result of the detection process can be found approximately with the expression

$$f = F_s \frac{k}{N}$$

In order for the estimated frequency value not to be too far from the original frequency value, a fraction of the total number of samples taken by the receiver is chosen to minimize the amount of mean square error between the estimated frequencies and the original frequency. In our example, on the receiver side, it is sufficient to use $N_t = 205$ samples out of $N = 800$ received samples to make an accurate estimation.

In the example, we considered the transmission of digit "1." The detection of other digits can be achieved in a similar manner. In the calculation of the discrete Fourier transform, the "k" values range from 0 to $(N-1)$. In the Goertzel algorithm, only some "k" values are used, as we mentioned in the previous pages. For sequences of small length, the Goertzel algorithm is more efficient, whereas, for longer sequences, the fast Fourier transform algorithm is more effective.

8.4   Fast Fourier Transform (FFT)

The method used in the development of the fast Fourier transform algorithm is based on some properties of the complex exponential signal. The complex exponential signal used in the Fourier transform is $e_N = e^{-j\frac{2\pi}{N}}$ from which we can write $e_N^{kn} = e^{-j\frac{2\pi}{N}kn}$.

The period of the complex exponential signal is N, that is, $e_N^{kn} = e^{-j\frac{2\pi}{N}kn}$. If $f(k) = e_N^{kn}$, then we have $f(k) = f(k+N)$.

Example 8.1 Find the period of $g(k) = e^{-j\frac{\pi}{3}kn}$.

Solution 8.1 The expression $e^{-j\frac{\pi}{3}kn}$ can be written as $e^{-j\frac{2\pi}{6}kn}$ from which the period is found as 6.

Reducing the Computational Complexity of the Discrete Fourier Transform
Discrete Fourier transform is calculated using

$$X[k] = \frac{1}{\sqrt{N}} \sum_{n=0}^{N-1} x[n]e_N^{kn}, 0 \le k < N-1 \tag{8.31}$$

where there are N^2 multiplication and addition operations.

Our aim is to do less multiplication and addition while calculating (8.31). Less computational complexity allows the calculations to be done faster by the processors and causes an increase in the data rate. Two different methods can be followed to decrease the computation amount. The first method is called decimation in time, and the second is called decimation in frequency. The computational gain of these two methods is the same. In this section, we will only explain the decimation in time method.

Decimation in Time
We will make use of the exponential signal e_N^{kn} to reduce the number of operations in the calculation of (8.31). Let us examine the period of the exponential signal for the cases $k = 2m$ and $k = 2m + 1$; for $k = 2m$ we have

$$e_N^{2mn} = e_{\frac{N}{2}}^{mn} \tag{8.32}$$

where it is seen that for $k = 2m$, the period of e_N^{kn} is $\frac{N}{2}$. For $k = 2m + 1$, we have

$$e_N^{(2m+1)n} = e_N^n e_{\frac{N}{2}}^{mn} \tag{8.33}$$

from which it is seen that for $k = 2m + 1$, the period of e_N^{kn} is $\frac{N}{2}$. Equation (8.31) can be calculated considering the odd and even values of k as

$$
\begin{aligned}
X[k] &= \frac{1}{\sqrt{N}} \sum_{n=0}^{N=1} x[n] e_N^{kn}, 0 \le k \le N - 1 \\
&= \frac{1}{\sqrt{N}} \sum_{m=0}^{\frac{N}{2}-1} x[2m] e_N^{k2m} + \frac{1}{\sqrt{N}} \sum_{m=0}^{\frac{N}{2}-1} x[2m+1] e_N^{k(2m+1)} \\
&= \underbrace{\frac{1}{\sqrt{N}} \sum_{m=0}^{\frac{N}{2}-1} x[2m] e_{\frac{N}{2}}^{km}}_{X_a[k]} + e_N^k \underbrace{\frac{1}{\sqrt{N}} \sum_{m=0}^{\frac{N}{2}-1} x[2m+1] e_{\frac{N}{2}}^{km}}_{X_b[k]}
\end{aligned}
\tag{8.34}
$$

In (8.34), the periods of $X_a[k]$ and $X_b[k]$ are equal to each other, and it is $\frac{N}{2}$. There are $\left(\frac{N}{2}\right)^2$ multiplication and addition operations for each term of (8.34). The total number of multiplication or addition operations is

$$\left(\frac{N}{2}\right)^2 + \left(\frac{N}{2}\right)^2 = 2\left(\frac{N}{2}\right)^2$$

Besides, for the calculation of $e_N^k X_b[k]$, additional N multiplication and addition operations are required. Thus, for the calculation of $X[k]$, the total number of multiplication or addition operations is

$$N + 2\left(\frac{N}{2}\right)^2$$

The logic used to separate $X[k]$ into two parts can be used for the signals $X_a[k]$ and $X_b[k]$. In other words, $\frac{N}{2}$ point discrete Fourier transforms can be written as the sum of two $\frac{N}{4}$ point Fourier transforms.

Separation of $X_a[k]$ into two parts can be performed as in

$$
\begin{aligned}
X_a[k] &= \frac{1}{\sqrt{N}} \sum_{n=0}^{\frac{N}{2}-1} x_a[n] e_N^{kn}, 0 \leq k < \frac{N}{2} - 1 \\
&= \underbrace{\frac{1}{\sqrt{N}} \sum_{m=0}^{\frac{N}{4}-1} x_a[2m] e_{\frac{N}{2}}^{km}}_{X_{a1}[k]} + e_N^k \underbrace{\frac{1}{\sqrt{N}} \sum_{m=0}^{\frac{N}{4}-1} x_a[2m+1] e_{\frac{N}{2}}^{km}}_{X_{a2}[k]}
\end{aligned}
\tag{8.35}
$$

Similarly $X_b[k]$ can be written as the sum of two terms as in

$$
\begin{aligned}
X_b[k] &= \frac{1}{\sqrt{N}} \sum_{n=0}^{\frac{N}{2}-1} x_b[n] e_N^{kn}, 0 \leq k \leq \frac{N}{2} - 1 \\
&= \underbrace{\frac{1}{\sqrt{N}} \sum_{m=0}^{\frac{N}{4}-1} x_b[2m] e_{\frac{N}{2}}^{km}}_{X_{b1}[k]} + e_N^k \underbrace{\frac{1}{\sqrt{N}} \cdot \sum_{m=0}^{\frac{N}{4}-1} x_b[2m+1] e_{\frac{N}{2}}^{km}}_{X_{b2}[k]}
\end{aligned}
\tag{8.36}
$$

That is, $\frac{N}{2}$ point discrete Fourier transform can be written as the sum of two $\frac{N}{4}$ point discrete Fourier transforms, and the number of multiplication and addition operations happens to be $N + N + 4\left(\frac{N}{4}\right)^2$. If the procedure is repeated until we get 2-point discrete Fourier transforms, the number of multiplication and addition operations for $N = 2^v$ can be calculated as

$$
\underbrace{N + N + \ldots + N}_{v-1 \; adet} + \underbrace{2^{v-1}\left(\frac{N}{2^{v-1}}\right)^2}_{=N} = N \log N
\tag{8.37}
$$

The recursive operation for the calculation of N-point discrete Fourier transform using 2-point DFTs is illustrated in Fig. 8.9.

Two-Point Discrete Fourier Transform

Since N-point Fourier transform is written as the sum of 2-point Fourier transforms, let us first obtain the flow graph of the 2-point Fourier transform and, then, express the flow graph of the bigger point Fourier transforms using the flow graph of 2-point Fourier transforms. Two-point Fourier transform calculation can be performed using

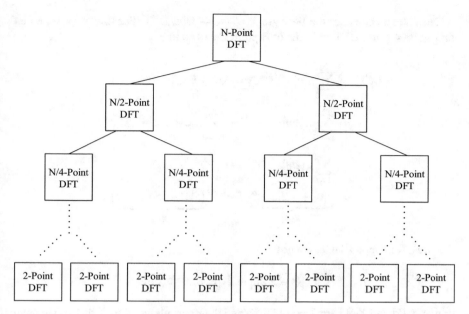

Fig. 8.9 Expressing N-point Fourier transform in terms of $\frac{N}{2}$-point Fourier transforms in a recursive manner

Fig. 8.10 Flow graph of
2-point DFT

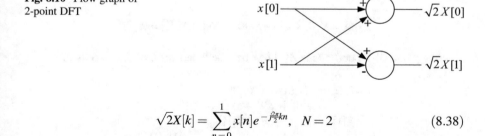

$$\sqrt{2}X[k] = \sum_{n=0}^{1} x[n]e^{-j\frac{2\pi}{2}kn}, \quad N=2 \tag{8.38}$$

which leads to the simple expression

$$\sqrt{2}X[k] = x[0] + x[1]e^{-j\pi k}, \quad k=0,1 \tag{8.39}$$

from which we get

$$\sqrt{2}X[0] = x[0] + x[1] \qquad \sqrt{2}X[1] = x[0] - x[1] \tag{8.40}$$

The flow graph of (8.10) can be drawn as in Fig. 8.10.

Now, let us consider the flow graph of the 4-point DFT. For this purpose, let us first write 4-point DFT in terms of 2-point DFTs as in

$$\sqrt{4}X[k] = \sum_{n=0}^{3} x[n]e^{-j\frac{2\pi}{4}kn}, N = 4, 0 \le k < 3$$

$$= \sum_{m=0}^{1} x[2m]e^{-j\frac{2\pi}{4}k2m} + \sum_{m=0}^{1} x[2m+1]e^{-j\frac{2\pi}{4}k(2m+1)}$$

$$= \sum_{m=0}^{1} x[2m]e^{-j\frac{2\pi}{4}k2m} + e^{-j\frac{2\pi}{4}k}\sum_{m=0}^{1} x[2m+1]e^{-j\frac{2\pi}{4}k2m}$$

$$= \underbrace{x[0] + x[2]e^{-j\pi k}}_{X_1[k]} + e^{-j\frac{\pi k}{2}}\underbrace{\left(x[1] + x[3]e^{-j\pi k}\right)}_{X_2[k]}$$

Thus, for 4-point DFT, we got

$$\sqrt{4}X[k] = X_1[k] = X_1[k] + e^{-j\frac{\pi k}{2}}X_2[k]$$

where $X_1[k]$ and $X_2[k]$ are 2-point DFTs and their periods equal to 2, that is, we have

$$X_1[k] = X_1[k+2] \rightarrow X_1[0] = X_1[2], X_1[1] = X_1[3]$$

and

$$X_2[k] = X_2[k+2] \rightarrow X_2[0] = X_2[2], X_2[1] = X_2[3]$$

Four-point DFT coefficients $X[k]$ can be calculated for $k = 0, \ldots, 3$ as in

$$
\begin{aligned}
\sqrt{4}X[0] &= X_1[0] + e^{0}X_2[0] \\
&= x[0] + x[2] + x[1] + x[3] \\
\sqrt{4}X[1] &= X_1[1] + e^{-j\frac{\pi}{2}}X_2[1] \\
&= x[0] + x[2]e^{-j\pi} + e^{-j\frac{\pi}{2}}(x[1] + x[3]e^{-j\pi}) \\
&= x[0] - x[2] + e^{-j\frac{\pi}{2}}(x[1] - x[3]) \\
\sqrt{4}X[2] &= X_1[2] + e^{-j\pi}X_2[2] \\
&= X_1[0] + e^{-j\pi}X_2[0] \\
&= x[0] + x[2] - (x[1] + x[3]) \\
\sqrt{4}X[3] &= X_1[3] + e^{-j\frac{\pi}{2}3}X_2[3] \\
&= X_1[1] + e^{-j\frac{\pi}{2}3}X_2[1] \\
&= x[0] - x[2] + e^{-j\frac{\pi}{2}3}(x[1] - x[3]) \\
&= x[0] - x[2] + e^{-j\frac{\pi}{2}}(x[1] - x[3])
\end{aligned}
\tag{8.41}
$$

Using the equations in (8.41), and Fig. 8.10, the flow graph of the 4-point DFT calculation can be drawn as in Fig. 8.11.

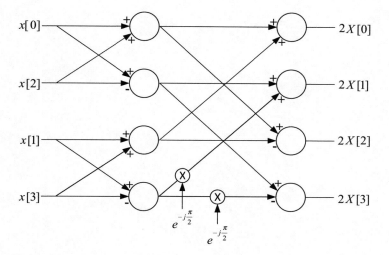

Fig. 8.11 Flow graph of 4-point DFT

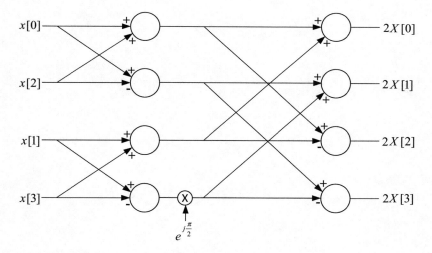

Fig. 8.12 Final flow graph of 4-point DFT

When Fig. 8.11 is inspected, we see that two separate multiplication operations with $e^{-j\frac{\pi}{2}}$ are performed. By modifying the graph as in Fig. 8.12, a single multiplication operation can be utilized.

Appendix A: Frequently Used Mathematical Expressions

$$\sin(\alpha \pm \beta) = \sin(\alpha)\cos(\beta) \pm \cos(\alpha)\sin(\beta) \tag{A.1}$$

$$\cos(\alpha \pm \beta) = \cos(\alpha)\cos(\beta) \mp \sin(\alpha)\sin(\beta) \tag{A.2}$$

$$\sin(\alpha)\sin(\beta) = \frac{1}{2}[\cos(\alpha - \beta) - \cos(\alpha + \beta)] \tag{A.3}$$

$$\cos(\alpha)\cos(\beta) = \frac{1}{2}[\cos(\alpha - \beta) + \cos(\alpha + \beta)] \tag{A.4}$$

$$\sin(\alpha)\cos(\beta) = \frac{1}{2}[\sin(\alpha - \beta) + \sin(\alpha + \beta)] \tag{A.5}$$

$$\sin^2(\alpha) + \cos^2(\alpha) = 1 \tag{A.6}$$

$$\cos^2(\alpha) = \frac{1 + \cos(2\alpha)}{2} \tag{A.7}$$

$$\sin^2(\alpha) = \frac{1 - \cos(2\alpha)}{2} \tag{A.8}$$

© The Editor(s) (if applicable) and The Author(s), under exclusive license to Springer
Nature Switzerland AG 2023
O. Gazi, *Principles of Signals and Systems*, https://doi.org/10.1007/978-3-031-17789-7

$$\cos(2\alpha) = \cos^2(\alpha) - \sin^2(\alpha) \qquad \text{(A.9)}$$

$$c = a + jb \rightarrow a = \text{Re}\{c\}, \quad b = \text{Im}\{c\},$$
$$c^* = a - jb \rightarrow a = \frac{c + c^*}{2}, \quad b = \frac{c - c^*}{2} \qquad \text{(A.10)}$$

$$e^{j\alpha} = \cos(\alpha) + j\sin(\alpha)$$
$$\sin(\alpha) = \frac{e^{j\alpha} - e^{-j\alpha}}{2j}, \quad \cos(\alpha) = \frac{e^{j\alpha} + e^{-j\alpha}}{2} \qquad \text{(A.11)}$$

$$\sum_{k=0}^{\infty} \alpha^k = \frac{1}{1-\alpha}, \quad |\alpha| < 1 \qquad \text{(A.12)}$$

$$[\cos(\alpha) + j\sin(\alpha)]^n = \cos(n\alpha) + j\sin(n\alpha) \qquad \text{(A.13)}$$

$$\sqrt[n]{re^{j\alpha}} = \sqrt[n]{r}e^{j\left(\frac{\alpha}{n} + k\frac{2\pi}{n}\right)}, \quad k = 0, \pm 1, \pm 2, \ldots \qquad \text{(A.14)}$$

$$e^{\alpha} = 1 + \alpha + \frac{\alpha^2}{2!} + \frac{\alpha^3}{3!} + \frac{\alpha^4}{4!} + \ldots \qquad \text{(A.15)}$$

$$\sin(\alpha) = \alpha - \frac{\alpha^3}{3!} + \frac{\alpha^5}{5!} - \frac{\alpha^7}{7!} + \ldots \qquad \text{(A.16)}$$

$$\cos(\alpha) = 1 - \frac{\alpha^2}{2!} + \frac{\alpha^4}{4!} - \frac{\alpha^6}{6!} + \ldots \qquad \text{(A.17)}$$

Appendix B: Fourier, Laplace, and Z Transform Pairs

Continuous-time Fourier transform and inverse Fourier transform formulas:

$$\widehat{f}(w) = \frac{1}{\sqrt{2\pi}} \int_{t=-\infty}^{\infty} f(t)e^{-jwt}dt \quad f(t) = \frac{1}{\sqrt{2\pi}} \int_{w=-\infty}^{\infty} \widehat{f}(w)e^{jwt}dw \qquad (B.1)$$

or

$$\widehat{f}(w) = \frac{1}{\sqrt{2\pi}} \int_{t=-\infty}^{\infty} f(t)e^{-jwt}dt \quad f(t) = \frac{1}{\sqrt{2\pi}} \int_{w=-\infty}^{\infty} \widehat{f}(w)e^{jwt}dw \qquad (B.2)$$

Continuous-time Fourier transform pairs:
The following pairs are obtained when (B.2) is used.

$$e^{jw_0t} \overset{\text{CTFT}}{\longleftrightarrow} \sqrt{2\pi}\delta(w-w_0) \qquad (B.3)$$

$$a>0, e^{-at^2} \overset{\text{CTFT}}{\longleftrightarrow} \frac{1}{\sqrt{2a}}e^{-\frac{w^2}{4a}} \qquad (B.4)$$

$$\cos(w_0t) \overset{\text{CTFT}}{\longleftrightarrow} \sqrt{\frac{\pi}{2}}[\delta(w-w_0) + \delta(w+w_0)] \qquad (B.5)$$

$$\sin(w_0t) \overset{\text{CTFT}}{\longleftrightarrow} j\sqrt{\frac{\pi}{2}}[\delta(w+w_0) - \delta(w-w_0)] \qquad (B.6)$$

© The Editor(s) (if applicable) and The Author(s), under exclusive license to Springer Nature Switzerland AG 2023
O. Gazi, *Principles of Signals and Systems*, https://doi.org/10.1007/978-3-031-17789-7

$$u(t) \overset{\text{CTFT}}{\longleftrightarrow} \frac{\sqrt{2\pi}}{2} \delta(w) + \frac{1}{\sqrt{2\pi}} \frac{1}{jw} \tag{B.7}$$

$$e^{-at}u(t) \overset{\text{CTFT}}{\longleftrightarrow} \frac{1}{\sqrt{2\pi}} \frac{1}{a+jw} \tag{B.8}$$

$$\frac{1}{2}[\delta(t-1) + \delta(t+1)] \overset{\text{CTFT}}{\longleftrightarrow} \frac{1}{\sqrt{2\pi}} \cos(w) \tag{B.9}$$

$$\delta(t) \overset{\text{CTFT}}{\longleftrightarrow} \frac{1}{\sqrt{2\pi}} \tag{B.10}$$

Discrete-time Fourier transform and inverse Fourier transform formulas:

$$F(\Omega) = \frac{1}{\sqrt{2\pi}} \sum_{n=-\infty}^{\infty} f[n]e^{-j\Omega n} \quad f[n] = \frac{1}{\sqrt{2\pi}} \int_{2\pi} F(\Omega)e^{j\Omega n} d\Omega \tag{B.11}$$

or

$$F(\Omega) = \sum_{n=-\infty}^{\infty} f[n]e^{-j\Omega n} \quad f[n] = \frac{1}{2\pi} \int_{2\pi} F(\Omega)e^{j\Omega n} d\Omega \tag{B.12}$$

Discrete-time Fourier transform pairs:
The following pairs are obtained when (B.11) is used.

$$\sqrt{\frac{2}{\pi}} \sin \frac{Wn}{n} \overset{\text{DTFT}}{\longleftrightarrow} \begin{cases} 1 & |\Omega| < W \\ 0 & W < |\Omega| < \pi \end{cases} \tag{B.13}$$

$$\frac{1}{\sqrt{2\pi}} \overset{\text{DTFT}}{\longleftrightarrow} \delta(t) \tag{B.14}$$

$$\alpha^n u[n], |\alpha| < 1 \overset{\text{DTFT}}{\longleftrightarrow} \frac{1}{\sqrt{2\pi}} \frac{1}{1 - \alpha e^{-j\Omega}} \tag{B.15}$$

$$\delta[n] \overset{\text{DTFT}}{\longleftrightarrow} \frac{1}{\sqrt{2\pi}} \tag{B.16}$$

$$e^{j\Omega_0 N} \overset{\text{DTFT}}{\longleftrightarrow} \sqrt{2\pi} \sum_{k=-\infty}^{\infty} \delta(\Omega - \Omega_0 - 2\pi k) \tag{B.17}$$

$$1 \overset{\text{DTFT}}{\longleftrightarrow} \sqrt{2\pi} \sum_{k=-\infty}^{\infty} \delta(\Omega - 2\pi k) \tag{B.18}$$

$$\delta[n - n_0] \overset{\text{DTFT}}{\longleftrightarrow} \frac{1}{\sqrt{2\pi}} e^{-j\Omega_0 n} \tag{B.19}$$

$$\sum_{k=-\infty}^{\infty} \delta[n - kN] \overset{\text{DTFT}}{\longleftrightarrow} \frac{\sqrt{2\pi}}{N} \sum_{k=-\infty}^{\infty} \delta\left(\Omega - \frac{2\pi}{N}k\right) \tag{B.20}$$

$$u[n] \overset{\text{DTFT}}{\longleftrightarrow} \frac{1}{\sqrt{2\pi}} \frac{1}{1 - e^{-j\Omega}} + \sqrt{\frac{\pi}{2}} \sum_{k=-\infty}^{\infty} \delta(\Omega - 2\pi k) \tag{B.21}$$

Laplace transform formulas:
Two-sided transform:

$$F(s) = \int_{t=-\infty}^{\infty} f(t)e^{-st}dt \tag{B.22}$$

Single-sided transform:

$$F(s) = \int_{t=0}^{\infty} f(t)e^{-st}dt \tag{B.23}$$

Laplace transform pairs:

$f(t)$	$F(s)$	ROC
$\delta(t)$	1	s plane
$u(t)$	$\frac{1}{s}$	$\text{Re}\{s\} > 0$
$-u(-t)$	$\frac{1}{s}$	$\text{Re}\{s\} < 0$
$e^{-at}u(t)$	$\frac{1}{s+a}$	$\text{Re}\{s\} > a$
$e^{-at}u(-t)$	$\frac{1}{s+a}$	$\text{Re}\{s\} < a$
$\cos(at)u(t)$	$\frac{s}{s^2+a^2}$	$\text{Re}\{s\} > 0$
$\sin(at)u(t)$	$\frac{a}{s^2+a^2}$	$\text{Re}\{s\} > 0$
$J_0(at)$	$\frac{1}{\sqrt{s^2+a^2}}$	$\text{Re}\{s\} > 0$

$$\frac{t^n}{n!}u(t) \qquad \frac{1}{s^{n+1}} \qquad \text{Re}\{s\} > 0$$

Z transform formula:

$$F(z) = \sum_{n=-\infty}^{\infty} f[n]z^{-n} \tag{B.24}$$

Z transform pairs:

$f[n]$	$F(z)$	ROC
$\delta[n]$	1	Entire z plane
$u[n]$	$\frac{1}{1-z^{-1}}$	$|z| > 1$
$-u[n-1]$	$\frac{1}{1-z^{-1}}$	$|z| < 1$
$\alpha^n u[n]$	$\frac{1}{1-\alpha z^{-1}}$	$|z| > |\alpha|$
$-\alpha^n u[-n-1]$	$\frac{1}{1-\alpha z^{-1}}$	$|z| < |\alpha|$
$n\alpha^n u[n]$	$\frac{\alpha z^{-1}}{(1-\alpha z^{-1})^2}$	$|z| > |\alpha|$
$-n\alpha^n u[-n-1]$	$\frac{\alpha z^{-1}}{(1-\alpha z^{-1})^2}$	$|z| < |\alpha|$
$\cos(\Omega_0 n)u[n]$	$\frac{1 - \cos(\Omega_0)z^{-1}}{1 - 2\cos(\Omega_0)z^{-1} + z^{-2}}$	$|z| > 1$
$\sin(\Omega_0 n)u[n]$	$\frac{1 - \sin(\Omega_0)z^{-1}}{1 - 2\cos(\Omega_0)z^{-1} + z^{-2}}$	$|z| > 1$
$\alpha^n \cos(\Omega_0 n)u[n]$	$\frac{1 - \alpha\cos(\Omega_0)z^{-1}}{1 - 2\alpha\cos(\Omega_0)z^{-1} + \alpha^2 z^{-2}}$	$|z| > \alpha$
$\alpha^n \sin(\Omega_0 n)u[n]$	$\frac{1 - \alpha\sin(\Omega_0)z^{-1}}{1 - 2\alpha\sin(\Omega_0)z^{-1} + \alpha^2 z^{-2}}$	$|z| > \alpha$

Bibliography

1. A. V. Oppenheim, A. S. Willsky with I. T. Young, Signals and Systems, Prentice Hall Signal Processing Series, 1983.
2. A. V. Oppenheim, R. W. Schafer, and J. R. Buck, Discrete Time Signal Processing, 2nd Edition, Prentice Hall, 1999.
3. R. A. Roberts, C. T. Mullis, Digital Signal Processing, Addison-Wesley, 1987.
4. D. W. Kammler, A First Course in Fourier Analysis, Prentice-Hall, 2000.
5. R. N. Bracewell, The Fourier Transform and Its Applications, 2nd Edition McGraw-Hill, 1978.
6. A. Papoulis, The Fourier Integral and Its Applications, McGraw-Hill, 1962.
7. J. G. Proakis, D. G. Manolakis, Digital Signal Processing: Principles, Algorithms, and Applications, 3rd Edition, Prentice Hall, 1995.
8. S. Haykin, B. V. Veen, Signals and Systems, 2nd Edition, Prentice Hall, 2003.
9. M. J. Roberts, Signals and Systems, McGraw-Hill, 2004.
10. E. Kreyszig, Advanced Engineering Mathematics, Wiley International Edition, 9th Edition, 2006.

Index

Printed in the United States
by Baker & Taylor Publisher Services